Education and Climate Change

Routledge Research in Education

Education and Climate Change

Living and Learning in Interesting Times

Edited by
Fumiyo Kagawa and David Selby

Routledge
Taylor & Francis Group
New York   London

First published 2010
by Routledge
711 Third Avenue, New York, NY 10017

Simultaneously published in the UK
by Routledge
2 Park Square, Milton Park, Abingdon, Oxon OX14 4RN

First issued in paperback 2012
Routledge is an imprint of the Taylor & Francis Group, an informa business

Library of Congress Cataloging in Publication Data

   Education and climate change : living and learning in interesting times / edited by
Fumiyo Kagawa and David Selby.
     p. cm.
   Includes bibliographical references and index.
  1. Education—Social aspects. 2. Education—Environmental aspects. 3. Climate
change. 4. Environmental education. I. Kagawa, Fumiyo. II. Selby, David.
   LC191.E268 2009
   370.11'5—dc22
   2009018473

ISBN13: 978-0-415-64915-5(pbk)
ISBN13: 978-0-415-80585-8(hbk)
ISBN13: 978-0-203-86639-9(ebk)

To Francis Robert Chapple, grandfather of David,
and Tomiko Kagawa, mother of Fumiyo,
who taught us to love and respect nature

Contents

Figures

Tables

Foreword

THE FATAL COMPLACENCY

Climate change is the greatest human-induced crisis facing the world today. It is totally indiscriminate of race, culture and religion. It affects every human being on the planet. But, so far, its impacts have fallen disproportionately.

In response to climate change, the word 'adaptation' has become part of standard vocabulary. But what does 'adaptation' mean? The answer to that question is different in different places. For most people in rich countries 'adaptation' has so far been a relatively painfree process. Cushioned by heating and cooling systems, they can 'adapt' to extreme weather with the flick of a thermostat. Confronted with the threat of floods, governments can protect the residents of London, Los Angeles and Tokyo with elaborate climate defence systems. In some countries of the North, climate change has even brought benign effects, such as a longer growing season for farmers.

Now consider what 'adaptation' means for the world's poorest and most vulnerable people—the 2.6 billion living on less than US$2 a day. How does an impoverished woman farmer in Malawi 'adapt' when more frequent droughts and less rainfall cut production? Perhaps by cutting already inadequate household nutrition, or by taking her children out of school? How does a slum dweller living beneath plastic sheets and corrugated tin in a slum in Manila or Port-au-Prince 'adapt' to the threat posed by more intense cyclones? And how are people living in the great deltas of the Ganges and the Mekong supposed to 'adapt' to the inundation of their homes and lands?

'Adaptation' is becoming a euphemism for social injustice on a global scale. While the citizens of the rich world are protected from harm, the poor, the vulnerable and the hungry are exposed to the harsh reality of climate change in their everyday lives. Put bluntly, the world's poor are being harmed through a problem that is not of their making. The footprint of the Malawian farmer or the Haitian slum dweller barely registers in the Earth's atmosphere.

What if dealing with climate change meant more than a flick of a switch to our friends in the industrialised world? Cushioned and cosseted, they have had the luxury of closing their minds to the real impact of what is happening in the fragile and precious atmosphere that surrounds the planet we live on. Where climate change has occurred in the industrialised world, the effects have so far been relatively benign. With the exception of events such as Hurricane Katrina in 2005, the inhabitants of North America and Europe have felt just a gentle caress from the winds of change.

Of course, rich countries can use their vast financial and technological resources to protect themselves against climate change, at least in the short-term—that is one of the privileges of wealth. But as climate change destroys livelihoods, displaces people and undermines entire social and economic systems, no country—however rich or powerful—will be immune to the consequences. In the long-run, the problems of the poor will arrive at the doorstep of the wealthy, as the climate crisis gives way to despair, anger and collective security threats. Failure to confront climate change head on amounts to a potentially fatal complacency.

Every child will recall the story of the three little pigs and the big bad wolf. In the world we live in, the bad wolf of climate change has already ransacked the straw house and the house made of sticks, and the inhabitants of both are knocking on the door of the brick house where the people of the developed world live. Our friends there should think about this the next time they reach for the thermostat switch. They should realize that while the problems of the Mozambican farmer might seem far away, it may not be long before their troubles wash up on their shores.

I very much welcome this collection of essays that speaks, first and foremost, to the part education can play in effecting climate change mitigation not 'adaptation'. The volume pulls no punches in its critique of the tepid response of school-age and higher education to the global warming challenge, a response that focuses on technological and scientific questions while more or less ignoring the ethics of the human condition, rampant consumerism and the global marketplace. The volume is very welcome, too, in its recurring emphasis on climate change as a global social justice issue and in its message that heart as much as head should guide what we do in the name of the human family.

Desmond Tutu
Archbishop Emeritus of Cape Town
Nobel Peace Laureate

Introduction

Fumiyo Kagawa and David Selby

For if these things are done when the wood is green, what will happen when it is dry?

Luke, 23

'May You Live in Interesting Times!' This allegedly ancient Chinese (but, some maintain, old Scottish) proverb invokes at one and the same time a curse and a blessing: a curse in that interesting times are ones of crisis, danger, and turbulence; a blessing in that periods of dislocation and uncertainty, laying bare the fragility of the normal and vulnerability of the taken-for-granted, are redolent with creative potential.

Climate change makes for most interesting times. There is widespread consensus among the international scientific community that human-induced climate change is happening and that, even with the most immediate and sustained mitigation efforts, there will be no escaping some time-lagged catastrophic consequences from past emissions of carbon dioxide and other gases.

In their most recent summary for policy makers, the 1000-strong international collectivity of scientists making up the physical science working group of the United Nations Intergovernmental Panel on Climate Change (IPCC) asserts that: "Warming of the climate system is unequivocal, as is now evident from observations of increases in global average air and ocean temperatures, widespread melting of snow and ice, and rising global average sea level" (IPCC 2007, 1). Confirming the anthropogenic nature of climate change and the likelihood of some "abrupt and irreversible" impacts (ibid. 13), the scientists draw upon a range of six low-to-high greenhouse gas emission computer simulations, as well as observed trends and improving understandings of positive feedback mechanisms within natural systems, to project a rise in surface air temperature of between 1.8°C (range 1.1°C to 2.9°C) and their best guess of 4.0°C (range 2.4°C to 6.4°C) during the twenty-first century relative to the 1980–99 period. Their estimate of sea level rise during the century, based on the same models, ranges from 0.18 meters to 0.59 meters relative to 1980–99, the figures being very conservative in that they do not factor in any future "rapid dynamical" Arctic and Antarctic ice conversion events (ibid. 7). Taking the melting of polar ice into account, current estimates by leading climate change scientists for sea level rises in the twenty-first century are in the range of 0.8 to 1.90 meters (Adam 2009a, 17). A poll of scientific experts from some twenty-six countries attending the March 2009 international climate change conference in Copenhagen reveals that 86% of those responding believe that global warming cannot be held within the bounds of the generally commonly expressed goal of restricting surface air temperature rise to 2.0°C, most believing a 4.0°C rise by 2100 is likely (Adam 2009b, 14).

The scientific climate change consensus somewhat breaks down over whether human society still has a window of opportunity to mitigate the more catastrophic impacts and, if so, how wide that window is. "We have a short period —very short period—in which to prevent the planet from shaking us off," writes George Monbiot (2006, 15), a view largely endorsed by the IPCC (2007, 20). Recently, James Hansen of NASA, a world-renowned climatologist, advised incoming US President Barack Obama that "we have only four years left to act on climate change" (McKie 2009, 44). Others are far less sanguine about there being any remaining window of opportunity. "The time," says leading climate change scientist Konrad Steffan, "is already five past midnight" (Kolbert 2007, 58). "Our future," writes James Lovelock (2006, 6), "is like that of passengers on a small pleasure boat above the Niagara Falls, not knowing that the engines are about to fail." For Joseph Romm, any perceived window of opportunity is narrowing fast. "Climate change is coming faster and rougher than scientists have expected," he warns (2007, 231). If anything marks changes in perception of climate change in the last ten years, it is the shift from seeing it as an incrementally and sedately unfolding, hence manageable, process to a fickle, unpredictable phenomenon, liable to become manifest in abrupt, cataclysmic, more or less unmanageable, bouts of accelerated impact (Lovelock 2009, 5).

Future histories, informed by their authors' meta-analyses of the plethora of scientific papers on climate change, paint a stark picture of the planetary future. Joseph Romm (2007) divides the twenty-first century into three eras: 2000–25 is when humans "reap the whirlwind," a period of increasingly extreme weather events, forewarning of a time when such events may well become the norm, but also a time of last opportunity for massive efforts, including sustained and draconian government intervention, to cut global carbon emissions. Should those efforts fall short, Romm sees 2025–50 as an era of "planetary purgatory," when the "point of no return" is reached and positive feedback loops kick in, with rising temperatures triggering the release of carbon stored in the tundra and forests. He calls 2050–2100 the era of "hell and high water," when, if carbon emissions have not been significantly stemmed by 2025, there will be a "tripled-CO^2 world" (ibid. 87), marked by massive sea level rises, and abandonment of lower-lying coastal areas as well as depopulation of inland continental areas decimated by drought and consequent aridity and seasonally recurring wildfire.

In *Six Degrees: Our Future on a Hotter Planet* (2007), Mark Lynas explores the environmental and human social impacts of rising surface air temperatures, devoting a chapter to each 1°C rise above the average temperature at the end of the last century. A one-degree warmer world would, among other things, reawaken the "slumbering desert" under the US and

Canadian wheat belt, push to breaking point the temperature resilience of the Amazon, rapidly accelerate Arctic meltdown, as well as intensify the global biodiversity crisis (Lynas 2007, 3–53). A two-degree warmer world would raise acidity in many ocean areas to levels toxic to sea creatures, and bring frequent repetitions of the European summer of 2003 with tens of thousands of human deaths per summer as well as agricultural devastation. A third of all species would be lost by 2050, presaging a "Silent Summer . . . devoid of birdsong, insect hum, and all the weird and wonderful noises that subconsciously keep us company" (ibid. 105). At three degrees, the Amazonian ecosystem would die and burn, Arctic ice would have almost disappeared, and high sea levels linked to storm surges would overcome all coastal cities where flood-protection investment had been inadequate (it would not be financially feasible to protect all cities, so stark choices would have to be made). "With structural famine gripping much of the subtropics, hundreds of millions of people will have only one choice left other than death for themselves and their families: they will have to pack up their belongings and leave. . . . Conflicts will inevitably erupt as these numerous climate refugees spill into already densely populated areas" (ibid. 171). At four degrees, the West Antarctic Ice Sheet becomes vulnerable, its collapse triggering huge sea level rises and seawater penetration of continental interiors. Much of the Mediterranean would be abandoned as environmentally inhospitable, "their residents flocking north, to overcrowded refuges in the Baltic, Scandinavia and the British Isles" (ibid. 192). At five degrees, "zones of uninhabitability"— areas where "large-scale, developed human society would no longer be sustainable"—would stand in stark contrast to "belts of habitability" contracting towards each pole, marking a new era of "enforced localism" (ibid. 224–25). The pressure from displaced populations on "zones of habitability" would be enormous, almost certainly occasioning violent conflict, tribalism, and the ever-lurking threat of genocide (ibid. 229). Dante's *Inferno* provides the inspiration, if that is the appropriate word, for Lynas' vision of a six-degree hotter world (ibid. 233, 256).

The chief scientist to the British government, John Beddington, has gone on record as anticipating the peak of a "perfect storm" brought about by a confluence of climate-induced crises around 2030. "If we don't address this," he warns, "we can expect major destabilization, an increase in rioting and potentially significant problems with international migration, as people move out to avoid food and water shortages" (Sample 2009, 14). Picking up on the same metaphor, Stephen Gardiner (2008, 25–42) writes of the "perfect moral storm" involving the convergence of a number of factors that threaten our ability to behave ethically in the face of looming climate change and conspiring to lead successive generations away from the path of acting according to intergenerational justice. Among these factors, he lists self-deception, selective attention to more "comfortable" aspects of the problem, and hypocrisy (ibid. 36).

Taking many forms and finding many expressions, denial has become a ubiquitous response to climate change. It was manifest in the deliberations of the G20 world summit in April 2009, where political leaders strove to reconfigure the old global economic growth model, arguably the root cause of the climate crisis, rather than explore a low-carbon global economy (Adam 2009b, 14). It is manifest in the framing of climate change exclusively or primarily as an issue calling for a scientific and technical fix rather than as a pathology of an ethically numb, inequitable, and denatured human condition (McIntosh, 2008). "The emphasis," Bob Doppelt asserts, "needs to shift from the biophysical sciences now to the social sciences if we have any hope of solving this problem" (Adam 2009b 14). "This is a moral moment," insists Al Gore (2007, 212).

Denial is also manifest in the failure of those in the metaphorical North of the planet to heed the issue of global climate injustice, whereby the impact of their polluting of the global atmospheric commons falls most severely on the least resilient populations of the South who have contributed least to the problem (Narain 2009, 10–11; Shiva 2008, 9–47). It is also ubiquitously evident in the "climate change equivalent of Orwellian Doublethink" (McIntosh 2008, 88) endemic among populations of the affluent North. People affirm the importance of combating climate change but not at the expense of their SUV, their 'can't do without' vacation and flight, their consumerist splurges. In this regard, so-called "green idealists" cannot be excluded, recent UK research finding that those most aware of environmental issues took the longest and most frequent flights, often seeing them as "reward" for otherwise green behavior (Adam 2008, 6). Political parties "in most rich countries have already recognized this," claims George Monbiot (2006, 41–42). "They know we want tough targets, but that we also want the targets to be missed. They know that we will grumble about their failure to curb climate change, but that we will not take to the streets. They know that nobody ever rioted for austerity." Rather than peering into "the unthinkable abyss" (McIntosh 2008, 208) and faithfully following through on the lessons of what is seen, the tendency is to go with 'business as usual' green options, their questionable 'fitness for purpose' notwithstanding, whether this be buying the indulgence of a carbon offset before flying, opting for 'green' consumer products, green energy, or engaging in international carbon trading (Monbiot 2006, 170–88, 210–12; Lovelock 2009, 147).

If what has been described is the curse of 'interesting times,' the blessing of such times comes in the myriad opportunities for rethinking, reorienting, and renewal. "We are now at a true fork in the road," says Al Gore (2007, 213–14). "It gives us an opportunity to experience something that few generations ever have the privilege of knowing: a common moral purpose compelling enough to lift us above our limitations." Rather than shying away from looming runaway climate change, the learning moment can be seized to think about what really and profoundly matters, to collectively envision a better future, and then to become practical visionaries in realizing that future.

At such a moment of enormous human challenge, formal, nonformal, and informal education have a potentially crucial role to play. In both school age and adult learning communities, learners of all ages can be invited to take up the challenge of understanding and rethinking the world, of shattering assumptions, shibboleths and the taken-for-granted, of deliberating where to go at this critical fork in the road. The learning process needs to have personal and societal transformative potential, flowing directly and naturally into direct community engagement. It should amount to an exercise in "anticipatory democracy" (Toffler 1971). Harold Glasser's (2007) concept of "active social learning" is relevant here. In contrast with "passive social learning" that is reliant on received wisdom and unquestioning embrace of orthodox assumptions and paradigms, active social learning is interactive, participatory, challenging and risky and, hence, harbors richer potential for emergence and transformation (ibid. 51–52). Glasser writes:

> Active social learning can take place in the context of a conversation, a course employing the Socratic method, dancing with a partner, symphony practice, a community meeting, an open, participatory review process and, although less visceral, video conferencing over the internet. (ibid. 51)

At the moment, the educational response to climate change has not been of this kind but has tended to mirror the response of society at large. The curriculum focus has been on imparting the science, but less often wrestling with the ethics, of global warming. The exhortation has been for personal change of a reformist (rather than transformative) nature. There has been an overarching absorption with the technological fix with a focus on reducing carbon emissions by the educational institution in question as well as by society at large (World Wide Fund for Nature UK, n. d.; Rappaport and Hammond Creighton 2007). Climate change learning experiences have, thus, tended to be confined within 'business as usual' parameters. There has been minimal recognition of the need to engage learners in openly debating and discussing the roots, personal meanings, and societal implications of climate change scenarios that are likely to play out during their lifetimes, and what needs to be done and achieved of a transformative nature by way of mitigation. The academy has tended to fiddle while Rome begins to burn.

The eight female and nine male contributors to this collection of essays, drawn from Africa, Asia, Europe, North America, Oceania, and Latin America, were asked to bring their insights, wisdom, and creativity to the question of what more creatively and critically transformative role education might play in helping mitigate looming and actual rampant and runaway climate change and its worst effects. Each contributor is a well-known figure in a particular field of social, political, and moral education, the fields represented being: adult education, antiracist education, citizenship education, education for sustainable development, emergency education, environmental education, faith and interfaith education, health education, Montessori-based education, peace education, school improvement education, and social justice education. In responding to the overarching challenge, each contributor was also asked to describe and reflect upon how their field was so far responding to the climate change issue, how the field might respond in increasingly 'interesting times' as runaway climate change sets in, and how its discourse, theory, and practice might in consequence change. Contributors were also challenged to think about what might happen to the balance of provision between formal, nonformal, and informal social learning during an unprecedented state of prolonged turmoil.

Contributors were additionally asked to close their piece with one or two concrete scenarios set within the near or more distant future and depicting social learning happening within contexts where the climate change threat is being somewhat ameliorated, and/or anticipated impacts of runaway climate change are well in evidence. We are grateful to them for their readiness to take up an envisioning challenge that does not come easy to most academics but which, given the endemic capacity to disregard the very real challenges of climate change, often by "burying bad news" (Lovelock 2009, 6), we felt to be necessary.

Edgar González-Gaudiano and *Pablo Meira-Cartea* offer a comprehensive and critical overview of barriers to promoting social change in the light of the complexity of the climate crisis. They problematize the ways in which scientific knowlege about climate change has been legitimatized, transmitted to, and perceived by the genearal public, ways which tend to ignore the structural nature of the problem and "obscure specific responsibilities" without questioning the dominant economic model and the prevailing Western lifestyle based on consumerism. Examining the issues of climate change from the prespective of social representation, the authors propose pedagogical and communication strategies predicated on social experience or "the everyday representation of the world" to encourage citizens' action and involvement.

In critiquing the response of the field of education for sustainable development to actual and looming climate change, *David Selby* identifies a prevailing 'business as usual' disposition that rather shirks away from critical exploration of the current economic growth model, marketplace globalization, and rampant consumerism. He also finds denial and cognitive dissonace among the field's proponents, and not least in their characterization of climate change as requiring an essentially scientific, technological, and management, rather than ethical and eco-spiritual, response. Proposing 'education for sustainable contraction' and 'education for sustainable moderation' as alternatives to the growth and market-tainted concept of 'education for development' with the potential to shift and deepen climate change discussion, Selby reviews key dimensions of the two conceptions.

Magnus Haavelsrud argues that, in a climate-changed world where natural conditions are increasingly imposing ever more constraining social, economic, political, and cultural limits, there is a need to build a new universalism, offering a dynamic vision of commonality, through peace learning. Peace learning as praxis he characterizes as transformative, learner-oriented, intersubjective, problem-centered, and diachronic (taking on board a longitudinal perspective). The discussion of the chapter focuses on the "synchronicity along the *space* dimension" in identifying and working from actionable "growth points" or dispositional points for transformative action for climate change: understanding and working with the dialectic relationship between macrorealities, on the one hand, and the microrealities of everyday life, on the other; creating exchanges between those reality levels ("the synergetic effects of selecting growth points at different levels on the space axis"). The proposed new universalism is conceived of as a product of a dialogical encounter between the multitude of voices making up indigenous knowledge systems, an encounter that, also through engagement with mainstream scientific knowledge, will validate subjugated knowledge systems and, in so doing, achieve cognitive justice.

For *Heila Lotz-Sisitka*, global dialog is a vital complementary dimension to contextualized and localized responses to climate change. Without such a dialog contextualized approaches "may run the risk of becoming bounded by conservative forms of contextualism, while seeking to be emancipatory." Drawing upon the leading-edge social justice theories, the chapter suggests a new concept of justice and discusses its implications for learning. They include "reflexive justice" guided by the "discourses of sufficiency and equity," "learning from" the environmentalism of the poor, and a "methodological cosmopolitanism" that addresses and incorporates mindfulness of the global situation and attendant risks while advancing "global and transboundary dialogs." Her chapter suggests a climate change educational framework should be "simultaneously critically transnational and globally reflexive while also supporting situated social learning processes that are contextually located and oriented towards agency, capability and risk negotiation in the everyday."

Guided by the "key questions of power, social difference, equality and justice" inherent to antiracist education, *George Dei* challenges and calls for a critical interrogation of dominant Western (or Eurocentric) notions of environment, development, and sustainability in considering the global challenge posed by climate change. Using the case of the Akosombo hydroelectric project in Ghana, where the tensions of 'tradition' and 'modernity' manifest themselves, he reflects on its pedagogical implications arguing that "we must consider Indigenous spirituality (cosmovision), and embodied knowledges as starting points, as operating central to the present day development/environment practice."

The field of emergency education, that is education responding to humanitarian crisis situations, already has considerable experience of practice within contexts that are, arguably, directly or indirectly triggered by climate change. *Fumiyo Kagawa*'s chapter points out the limitations of current narrow and linear conceptualization of emergencies and the danger of educational responses merely perpetuating neoliberal notions of development. The chapter argues the importance of enhancing resilience at the individual, social structural, and ecological levels predicated on alternative, human potential, and communitarian understandings of development. Considering the prospects of ever-more-severe climate change where emergencies increasingly become the norm, the chapter emphasizes the importance of education directly responding to trauma, learning to help both expand and deepen the scope of justice concern, and educational efforts directed towards sustaining a sense of hope among learners. It also argues for the critical importance of changing the direction of the flow of emergency-associated knowledge so that the flow is primarily from community rather than imposed on community.

In the light of climate change, *Ian Davies* and *James Pitt* first examine the nature of citizenship and the meaning of citizenship education, focusing on four areas: citizenship and location, expectations of and for citizenship, citizenship and crisis, and citizens and their perspective on the environment. Through a case study investigation of the National Curriculum of England, their chapter considers the relationship between citizenship education and education for sustainable development. The chapter highlights the overall lack of climate change concerns in citizenship education and the limited impact of education for sustainable development on the current National Curriculum. The authors propose "a broadly based conceptual approach that allows learners to explore environmental issues and develop knowledge, skills and dispositions that are needed for a sustainable future" with a particular emphasis on developing a right relationship with nature. As dire scenarios of climate change unfold, the authors stress the significance of redefining the notion of legal citizenship from an ethical point of view and foresee a possible shift from formal school learning to "community learning centers" in which the school/community division disappears as learners are encouraged to engage in social action. "Within this scenario," they write, "there is a humanization of education, a recognition that content, process and outcomes are in critical, dialogical and developing relationship with a developing understanding of humanity and its right relationship with nature."

Starting in the late 1970s, the field of school improvement involves both the study and practice of helping a school become a proactive, responsible, and autonomous organization. There are already a number of ways in which the field offers the social learning that will enable learners to address climate change: collegial school cultures, 'distributed' leadership, supporting staff, school self-evaluation, and student leadership and participation, to identify a few. However, in the face of climate change, *Jane Reed* identifies the need for "a larger sense of purpose, one where community transformation and environmental concern are connected to each other and to the purpose of education. The agenda includes people, their habitat and health, as well as their connection to the biosphere." More specifically, she highlights the importance of articulating and critiquing the dominant worldview underlying its activities, and developing an ecological worldview. In anticipating the disruption of conventional schooling in the event of rampant climate change, her chapter anticipates that "a move to more community-based learning will be a critical step," also emphasizing the importance of shifting systemic focus from performance, teaching, and attainment to learning. Change will require "laterality," a dialog between the currently divided silos of school improvement and sustainability education.

For *Darlene Clover* and *Budd Hall* the field of environmental adult education has been calling for a move away from an "expert information driven paradigm" and an "individual change, behavioral modification paradigm" towards "the idea of political empowerment, people's knowledge and collective action for change." The processes of critical and creative learning and engagement that the field promotes, they argue, are particularly important when corporate and governmental activities privileging economic growth and profit lie at the root of developments liable to aggravate climate change. Their chapter tells the story of educative-activist activities initiated by local women sewing quilts in response to a gas-field development on Vancouver Island, Canada. It is an example of how collective voices and energy for change can be effectively mobilized using aesthetic, creative, and participatory approaches. The "soft, gentle approach as well as the beauty of and color of the quilts" mobilized local communities, while grabbing the attention of the wider public. The story highlights the importance of validating "knowledge of so called ordinary people—what they know, think and feel about the issues in ways which are truly expressive."

Toh Swee-Hin and *Virginia Cawagas* describe the ecological wisdom deep within faith traditions that is being drawn upon within faith and interfaith contexts in response to the threat of climate change. They identify key assets that faith communities can bring to climate change discourse, while recognizing that those assets can be better capitalized upon by applying to them a transformative social learning paradigm based on principles of holism and interrelationship; participatory, horizontal, and creative dialog; values-based education; and critical empowerment. That application is happening in some faith-based institutions but by no means all. The authors offer insights into a range of faith-based and interfaith examples of community action on climate change; they discuss the greening of faith spaces; they outline faith-based and interfaith climate change advocacy within international arenas; they offer examples from many countries of innovative faith-based and interfaith formal curricula. While affirming these hopeful signs of progress, they join other authors by asking whether climate change solutions are "still framed within the dominant paradigm of economic 'progress' and 'development' " or whether they constitute a step towards "a distinctly alternative paradigm of how humanity needs to live in holistic and interdependent relationships with the earth and all beings and parts of the universe." Faiths, they argue, have a key role to play in such a transition.

Janet Richardson and *Margaret Wade* examine how health professionals, through policy and educational practice, can become equipped to deal with the implications of climate change for health and sustainability. The authors offer a brief summary of the consequences of climate change, as they are likely to impact upon the health of populations. They also consider the role of communication and education in facilitating behavioral change designed to marry healthier lifestyle with positive environmental impact. Communicating health information through education in ways that encourage motivation, build a sense of personal locus of control and agency, and generally engage and empower are explored. UK examples of grassroots action that supports and promotes local engagement are described, as well as policy directives that foster sustainability within and through health promotion initiatives.

Drawing upon her work in Montessori classrooms in very vivid ways, *Wendy Agnew* offers exciting examples of the facilitation of learning using art, dance, drama, symbol, music, poetry, outdoor experience, and student-designed research to build embodied earth connectedness and sensibility in children, her chapter in both format and style remaining as congruent as is possible within the confines of an academic book to her philosophy. She describes the learning as following "a cognitive old-growth forest model," as a "complex adaptive system," rather than "industrial monoculture" model. Such kinds of "impassioned intervention" are vital, she argues, in the face of the "avalanche of human folly" now becoming manifest in climate change.

The book closes by drawing together leitmotifs running through all or most of the twelve contributions by offering a synoptic *Critical Agenda for Interesting Times*.

REFERENCES

Adam, D. 2008. Green idealists fail to make the grade, says study. *Guardian*, September 24, 6.

Adam, D. 2009a. Sea level could rise more than a metre by 2100, say experts. *Guardian*, March 11, 17.

Adam, D. 2009b. Scientists fear worst on global warming. *Guardian*, April 14, 14.

Gardiner, S. 2008. A perfect moral storm: Climate change, intergenerational ethics, and the problem of corruption. In *Political theory and global climate change*, ed. S. Vanderheiden, 25–42. Cambridge, MA: MIT Press.

Glasser, H. 2007. Minding the gap: The role of social learning in linking our stated desire for a more sustainable world to our everyday actions and politics. In

Social learning towards a sustainable world: Perspectives, principles and praxis, ed. A. Wals, 35–62. Wageningen, The Netherlands: Wageningen Academic.

Gore, A. 2007. *The assault on reason*. London: Bloomsbury.

IPCC. 2007. *Summary for policymakers of the Synthesis Report of the IPCC Fourth Assessment Report*. Geneva: IPCC Secretariat.

Kolbert, E. 2007. *Field notes from a catastrophe: A frontline report on climate change*. London: Bloomsbury.

Lovelock, J. 2006. *The revenge of Gaia: Why the Earth is fighting back—and how we can still save humanity*. London: Allen Lane.

Lovelock, J. 2009. *The vanishing face of Gaia: A final warning*. London: Allen Lane.

Lynas. M. 2007. *Six degrees: Our future on a hotter planet*. London: Fourth Estate.

McIntosh, A. 2008. *Hell and high water: Climate change, hope and the human condition*. Edinburgh: Birlinn.

McKie, R. 2009. We have only four years left to act on climate change—America has to lead. *Observer*, January 18, 44.

Monbiot, G. 2006. *Heat: How to stop the planet from burning*. Toronto: Doubleday Canada.

Narain, S. 2009. A million mutinies. *New internationalist* 419 (January/February): 10–11.

Rappaport, A. and S. Hammond Creighton. 2007. *Degrees that matter: Climate change and the university*. Cambridge, MA: Massachusetts Institute of Technology.

Romm, J. 2007. *Hell and high water: Global warming—the solution and the politics—and what we should do*. New York: William Morrow.

Sample, I. 2009. 'Perfect storm' of scarcity will unleash global turmoil. *Guardian*, March 19, 14.

Shiva, V. 2008. *Soil not oil: Climate change, peak oil and food insecurity*. London: Zed Books.

Toffler, A. 1971. *Future shock*. London: Pan.

World Wide Fund for Nature UK. n. d. *Climate chaos: Learning for sustainability*. Godalming: WWF UK.

1 Climate Change Education and Communication

A Critical Perspective on Obstacles and Resistances[1]

Edgar González-Gaudiano and Pablo Meira-Cartea

OPENING IDEAS

Over the last few years, climate change has become an issue frequently addressed not only in the media but also in multiple areas of everyday life. From the point of view of the lay public, and with or without a basis in fact, many current events are now blamed on climate change, including food shortages and the consequent increase in food prices, mass migration from rural to urban regions and to developed countries, the growing vulnerability of coastal areas in the face of extreme weather events, and the spread of desertification, to mention just a few of these. The emergence of public and political interest in climate change has renewed the importance of environmental concerns in national and international political agendas, where their relative weight had declined since the 1992 UN Conference on Environment and Development in Brazil.

In addition to its recent emergence in the political and public arena, climate change is an extremely complex epistemological phenomenon. Not only does it cover a range of issues that have been studied individually and by separate scientific disciplines, but the connections between these have given rise to a whole new architecture of inquiry and set of challenges to conventional available knowledge. It is thus a hybrid theme essentially founded on uncertainty. Such uncertainty derives from the impossibility of controlling—or even identifying—all of the relevant variables, and to know how these are linked to each other. This is particularly true in relation to the goal of making predictions and moving from a global scale to a regional, subregional, or even local scale of knowledge on climate and climate change. As far as we know, according to the most recent assessment of the Intergovernmental Panel on Climate Change (IPCC 2007), climate change is real and has been caused, beyond reasonable doubt, by human activity. In the face of ever more conclusive evidence produced by climate change science, and despite the efforts of interest groups to obstruct international negotiations, progress has been made in furthering agreements and policy, in addition to the growth of environmental literacy within society as a whole.

We understand environmental literacy to mean not only the acquisition of information about the environment, but that such knowledge rests on a political and ethical substrate as well as a critical social practice linked to the concept of citizenship (González-Gaudiano 2007; Caride and Meira 2001). In contrast, the predominant notion of environmental literacy derives from a variant of environmental education initially promoted by UNESCO and UNEP, through their International Environmental Education Program (González-Gaudiano 2007), and in particular by its editorial series. This variant treated environmental education as an element of formal schooling. It was therefore viewed more as part of a curriculum than as a social process, and was very much focused on the natural sciences. In this chapter we will broach the implications of this inconvenient legacy for the development of educational programs to address climate change, and we will then discuss some of the main obstacles and social objections to the promotion of these programs.

ENVIRONMENTAL EDUCATION OR SCIENTIFIC LITERACY

Environmental education is a multidiscursive field. In addition to the environment per se, multiple theories and conceptions of education converge here, ranging from the most instrumental teaching approaches to the most critical perspectives and including a wide range of stances on environmental problems allied with both deep and social ecology.[2] We see it as a space of both struggle and opportunity for the reconstruction of a web of relations between humans and the environment and with each other. This web has faded with the advance of civilization, but could enable us to develop new treaties between cultures, societies, and nature and give rise to new values, languages, and meanings that could lead us responsibly to the social change that is so critical at this time. Thus, we do not share the view that environmental education should be promoted as separate from political and ethical content and focused exclusively on scientific-technical knowledge with the notion that this permits everyone to act autonomously according to their own principles and criteria (Aldrich and Kwong 1999). That is, we disagree with a vision of science as deus ex machina capable of solving complexity and uncertainty. Such a view depicts scientific knowledge as a mirror of reality, and hides the fact that science is part of a process of social construction and not a finished and infallible product. That said, and regardless of their objects of study, all sciences are social and, therefore, not so much subjected to an alleged objectivity but to an objectification that makes them intelligible, following normative and normalizing conventions established by specific scientific communities where nonscientific interests are also present. The problem is that science, understood as a value-neutral and aseptic (allegedly nonsubjective) product, has been the dominant approach in educational programs on climate change.

Within this view, the climate change crisis stems from a generalized lack of understanding of the energy and material flows between society and nature (Foladori 2000). Therefore, educational programs become processes of scientific literacy aimed at reinforcing the teaching of science and based in the transmission of scientific content, particularly of ecology (see Orr 1994 and Capra 2003).[3] For this reason, educational proposals related to climate change revolve around a series of issues relative to atmospheric composition, fluid dynamics, the basic principles of thermodynamics, the carbon cycle, the hydrogen economy, the oceans and the Gulf Stream, among many others (see Flannery 2006). The idea implicit in such an approach is that the problem is so complex and serious that only scientists and specialists know the answers, and thus that they are the only ones capable of defining what political actions should be taken.

This is not the end of it. Such a perspective starts from the premise that once people acquire updated and valid scientific information, their attitudes, values, and behavior will change and they will become environmentally literate citizens. This, however, is a simplistic, mechanistic, and deterministic assumption which has already been much discussed, not just in terms of the instructional approach (transmission of knowledge) (Sterling 1996), but also, above all, because of the precariousness of the results after years of implementation, as well as the undesirable side-effects observed (Sterling 2001).

Thus, a fundamental question is, what kind of social representation is in the best interest of society? Is it a 'scientific education or literacy' that tries to adjust social representations to the best available science or an 'environmental education' grounded in those aspects most significant in people's lives and as a means of motivating changes in lifestyles? For example, the social dimensions of climate change are as little known to Western citizens as the energy model (Meira and Arto-Blanco 2008).

In addition, scientific literacy is impregnating popular culture by way of intermediaries who are typically not scientists. More commonly it is transmitted by a wide range of spokespeople, including politicians, the news media, and communicators of science who 'bring' or 'explain' the scientific truths to the general public—that undifferentiated and scientifically illiterate mass that presumably requires scientific knowledge in order to live more environmentally friendly lives—and thereby displace common sense. This repetition of information on the climate change problem on television—the medium that many people depend on to guide their lives—becomes a new means of legitimating scientific knowledge through what Lyotard (1984) calls "performance," that is the "capacity to produce reality." In this way, a new meta-narrative, or total philosophy of history, takes shape, establishes ethical and political prescriptions for society, and tends to progressively regulate decision-making and determine what can be legitimately claimed as truth.

Thus, there are numerous climate change educational programs in circulation that provide recent facts, longitudinal studies, expert opinions, reports produced by prestigious institutions, and other sources. A social representation of this complex phenomena thus emerges that, in the end, rests on a small group of simplistic, atomized, and disarticulated actions, which are individual in nature and without a broader program. These measures fit within the range of possibilities based on the economic conditions of each person, from buying a hybrid car to changing from incandescent light bulbs to energy-saving lamps. Normally, this set of suggestions and recommendations for 'saving the world' is delivered from a perspective of general and undifferentiated responsibility: since we are all responsible for the problem, we are all equally responsible for the solution. Beyond a broad criticism of those who have failed to ratify the Kyoto Protocol, such rhetoric obscures specific responsibilities, the extravagance of our way of life is not questioned, nor is the style of growth based on consumerism. This is particularly true in periods of international financial crisis, whose 'solutions' rely on a dynamization of the market.

This raises question such as: Do we need to know the carbon cycle and the laws of thermodynamics in order to accept the need to change the type of light bulb used? Are these simplistic measures substitutes for structural actions that would genuinely allow us to move toward the kind of change that is required and so often postponed?

Before moving to a more detailed analysis of the complexity implicit in the implementation of educational programs for climate change, we would like to make a few points to clarify our position. Nothing is further from our intentions than to suggest that people should stop the specific actions that they are beginning to take to reduce high levels of energy consumption, particularly because electricity generating plants are the single largest source of greenhouse gases (although many people think that electricity is clean because it does not produce emissions during consumption). Millions of individual actions obviously can have significant effects on the climate system. What we are against is the transmission of an implicit idea that through these actions the problem is solved, and thus individual responsibility is exhausted. The substitution of all the incandescent bulbs in the houses of the developed world will not resolve the problem unless measures are taken in the diverse areas that now constitute the Western lifestyle. Neither are we opposed to the diffusion of available scientific knowledge. Today it is one of the most valuable resources for understanding to some extent the atmospheric changes that are occurring and their consequences. But this does not imply that broad scientific literacy will automatically trigger the change in behaviors, habits, and values expressed in peoples' everyday lives. At the very least this belief should be considered naive. Rather, we need a pedagogical strategy that is based on scientific knowledge and social experience among other factors, that establishes organized collective action toward clear ends and, above all, that challenges the normative values that organize life in society and is oriented to weaken the resistances, and the cognitive, psychosocial, and cultural barriers that impede change.

Only in this way will individual action become meaningful and contribute to overcoming the current state of things.

THE REPRESENTATION OF CLIMATE CHANGE IN 'POPULAR CULTURE'

Climate change has a series of characteristics that make it a fundamentally complex object, both from the scientific viewpoint, as well as from that of popular culture. Climate science has deepened its knowledge through the multidisciplinary and prolonged collection, analysis, and interpretation of direct and indirect evidence, in a process subject to permanent debate, negotiations, and controversy. This process has occurred fundamentally within the framework of the IPCC and of the negotiation processes that originated in the United Nations Framework Convention on Climate Change. The IPCC's reports are thus generally considered to be the principal scientific reference, given the broad consensus in their development and the independence from spurious interests that has characterized the scientists responsible for their preparation. The complexity of climate change is expressed also in the socioeconomic and political arena. The need to take decisions with the urgency required by this threat runs up against the dominant model of development sustained by fossil fuel energy sources that have become indispensable for our mode of exploitation, production, and consumption. It is practically impossible to conceive of or to apply structural policies and programs that mitigate the effects of climate change without affecting central aspects of the current economic model and the sociocultural order tied to it.

It is probably the first global environmental problem that is radically systemic: practically all ecological and human systems are implicated and are being or will be affected by its consequences in the short, medium, or long term. The 'solutions,' whatever the strategic definitions adopted (mitigation or adaptation), require a fundamental change in the established means of transforming, distributing, and consuming energy in order to significantly reduce anthropogenic greenhouse gas emissions, as well as the preservation and promotion of natural carbon storage and sinks. The necessary direction of change has been identified, but many resistances and social inertia hinder this route. One of these forms of resistance is the social representation of climate change, that is, the way in which human societies, particularly the most advanced, are culturally constructing their representations of climate change. This process not only involves science and its intermediaries, as mentioned above. Cultural production that responds to an 'epistemology of common sense' also plays a fundamental role in the social representation of climate change. In the construction of this representation, the transfer of scientific knowledge to citizens as part of 'environmental literacy' plays a role, but there are also other processes and factors outside the logic of scientific knowledge production that we seek to briefly identify. Some of these have to do with the 'complex nature of the problem,' and others with the logic of 'common sense,' and others more with the reelaboration and reinterpretation of knowledge when it becomes part of socially shared knowledge.

Consideration of the roots of the threat is essential in order to correctly focus the educational and communicational challenge presented by climate change. In fact, the main 'cultural barrier' is the structural nature of the problem. On this basis we can identify the various cultural, sociocognitive, and psychosocial obstacles that condition the social representation of the problem and hinder the adoption of the changes in individual and collective lifestyles that are most related to activities that unbalance the climate. For the purposes of categorization and systematization we have grouped these in three broad areas:

1.   those derived from the complex nature of the problem;
2.   those emerging from the moral and sociopolitical implications; and
3.   those related to the psychosocial and cognitive processes that condition the representation of climate change.

The Difficulty Complexity Poses to Social Understanding

Change in Climate: Natural or Human Factors?

One of the difficulties for public comprehension of climate change is that both human and natural factors (which have been the climate-transforming forces in the past) are at work. The IPCC's reports include natural factors in their explanatory models for climate change and consider them to have had a definitive effect in the abrupt climate transformations of the past. This possibility feeds the arguments of the "climate change deniers" who question the existence of climate change as well as the policies to address it: if this is of natural etiology then such policies would make no scientific or economic sense and an adaptive approach should be adopted. The IPCC's *Fourth Assessment Report* (2007) includes a detailed analysis of the climate system's energy balance, and weighes the influence of factors such as the increase in concentrations of greenhouse gases, terrestrial albedo, the role of aerosols, and fluctuations in solar radiation in the changes leading to global warming. Even including natural processes, for the IPCC it is clear that the largest share of observed climate "forcing" is attributable to human action.

Causes of Climate Change: An Aggregate Effect

At the present moment, individual and local actions can no longer be understood as good or bad per se since they contribute to global effects that are often synergistic and form part of unpredictable causal chains. A single and apparently innocuous action, like turning on a fan, can become danger-ous if repeated simultaneously in millions of homes. The climate change problem stems from the accumulation of greenhouse gas emissions from numerous individual contributions, whose 'share,' when considered indi-vidually, appears irrelevant at the atmospheric scale, but which, in aggre-gate, substantially alter the climate system. An evaluation of the adequacy of a single action is increasingly difficult and, beyond the 'here' and 'now,' depends on progressively more complex and difficult to comprehend con-textual situations

Temporal and Spatial Dimensions of Climate Change

Climate change is differentiated in time and ubiquitous in space. From the temporal point of view, the velocity of change over the last few decades is extremely rapid compared to the natural evolution of the climate. Nonethe-less, at the scale of a human life, the change—starting with the Industrial Revolution and whose critical manifestations are projected for the second half of the twenty-first century—is perceived as 'slow.' The asynchrony between the 'rhythm' of climate change and the personal biographical 'rhythm' of humans impedes recognition of the problem as well as a per-ception of urgency in addressing it.

- On the one hand, this is true because there is a time lag between the emission of greenhouse gases and their (differentiated and accumula-tive) effects on the atmosphere and this lag impedes an understanding of the relationship between actual emissions and their environmental consequences.

- On the other, it is difficult to realize the urgency of change since nei-ther the negative effects of climate change, nor the positive effects of actions aimed at reducing emissions, are perceived instantaneously and the inertia in the system is also not understood.

The difficulty in seeing and assigning a value to the 'rhythm' and 'progress' of climate change is an obstacle to understanding a process that is 'slow' by human perception, but too fast from the perspective of climate and ecology. It is not easy to understand that the impact of climate change and the capac-ity for human and ecosystem adaptation depend on the *rate* of change. This variable means that different human communities have different vulnerabili-ties, which is also true of species whose adaptation capacities tend to operate over long periods of time that facilitate the emergence of adaptations.

To these difficulties we have to add the possibility that transformations will occur through 'threshold' effects, that is, progressive and accumulative changes that are nearly imperceptible, then provoke sudden and sharp shifts from the established equilibrium. Paradoxically, since the consequences of extreme meteorological events tend to have major emotional impact (e.g., Hurricane Katrina), human beings tend to overrate these but to undervalue the subtle but important changes in our daily environment (e.g., the increase by tenths of a degree in average temperatures or changes in phenological calendars).

Another obstacle related to temporal perception is the difficulty in understanding the irreversability and inertia of climate change. A lack of in-depth understanding of the problem—from the scientific viewpoint—among segments of the population that have undervalued the risks or question its existence functions as an excuse to postpone the adoption of radical changes. Nonetheless, climate science warns that the longer we delay measures to reduce greenhouse gas emissions, the deeper and more irreversible climate change consequences will be in the long term, both for the biophysical environment and human communities.

From the spatial point of view, the ubiquity of climate change also generates problems for its social representation. It is not easy—even for science—to determine how climate change is affecting in the present, or will affect in the future, each region and place on the planet. The global predictions in the models and the scenarios used by the IPCC lose reliability at finer geographic scales, adding additional levels of uncertainty to social perceptions of the problem. Moreover, the fact that the effects are not the same across different zones on earth contrasts with the commonly transmitted idea that associates climate change with an average increase in the temperature throughout the entire planet. In fact, average temperatures at points on the terrestrial surface may reflect this increase or be above or even below this level. Global predictions, massively disseminated through the news media, can contradict local experience. For example, climate change is often associated with a reduction in the global availability of freshwater, but some human communities will suffer recurrent flooding that was previously unknown.

From this dual temporal and spatial perspective, climate change can appear to the population as an abstract and atemporal problem, as a 'counter-intuitive phenomenon' whose causes are difficult to determine and whose consequences will be felt on a scale beyond our immediate life horizon. If, for science, it is difficult to establish a linear, clear, and precise relationship between the problem identified as global, and its global, regional, and local expressions, for lay people this difficulty is multiplied.

Uncertainty: Scientific and Social Meanings

Uncertainty is an essential element of scientific knowledge which can only evolve through permanent review, controversy, and debate by the community that develops it. The uncertainty inherent in climate change knowledge complicates the relationships between climate science and the political and social spheres, and makes communication between science and society particularly difficult. Climate change is also a scientific object characterized by enormous complexity: it implies multiple systems and variables in synergistic relationships; it is addressed through different disciplines; there are important gaps related to fundamental aspects in its genesis; and efforts are being made to offer a prospective vision as to its evolution in the medium and long terms in order to orient political decision-making. If uncertainty is inherent to the scientific process, its projection into the social sphere can cause confusion, impede awareness as to the seriousness of the problem, and inhibit an understanding of the necessity of urgent action. The focus on the relative indeterminacy of knowledge suggests waiting for greater certainty, acts as a disincentive for preventative action, and feeds an historic optimism based on the naive belief in progress that predicts that the future will necessarily be better. The economy of effort that guides daily life finds in uncertainty a justification for postponing decision-making. Uncertainty in the scientific construction of climate change is transferred to society in different forms:

1. The scientific method requires that data and its interpretation be submitted to the critical review of a community of researchers in a methodological process that is fundamental to the collegial development of knowledge in all of the sciences. From the point of view of lay citizens though, the public echo of the controversies within the scientific debate can be interpreted as insecurity, confusion, or division within the core of the scientific community, thus feeding doubts as to the very existence of the problem, its causes, and its possible consequences, and to the attention that it merits as a real threat. Scientific uncertainty can be manipulated to feed 'climate change denial' and inaction.

2. In addition, the mass media—the main source of environmental information for many people—tend to dedicate the same time to scientific voices that alert us to the threat of climate change and the human role in its generation (the clear majority) as to the voices of scientific minorities, institutions, and 'opinion leaders' that deny the phenomena, attribute it to natural factors, do not consider it a threat, or that undervalue it through appeals to the adaptive capacities of ecosystems and societies (see Lomborg 2001).

3. Another factor that amplifies uncertainty is the tendency to highlight the gaps in climate knowledge relative to the certainties that have been identified and confirmed. It is important to take into account that the more we know about the climate and climate change, the more new gaps will be detected in our knowledge of the phenomena that require additional research.

4. The same effect is produced by the emphasis on the range of variation within the predictive models for the future climate relative to the increasing probability stated in the IPCC's recent report (2007) that the worst scenarios will become reality.

The Sociopolitical and Moral Expressions of Climate Change

From Scientific Complexity to Social Complexity

Climate change forms part of the process of globalization in three senses. First, it is one of the most emblematic 'products' of the generalization and universalization of the model of production originating in the Industrial Revolution. Second, it demonstrates that this model is in crisis, and this paradoxically derives from its very success as a way of organizing material production, as well as in the creation of a relatively safe and stable environment, and the generation of shared meanings for an important, if minority, portion of humanity. Third, the solutions, whatever they may be, require a high level of international consensus, since to be effective they must be globally assumed and applied. Thus, sociopolitical complexity overlaps with scientific complexity, or to push this even further, a synergistic effect between the two creates a hybrid where the scientific, the social, and the political are amalgamated and in permanent tension.

At the same time, the complexity that surrounds this threat is also expressed in moral terrain. Human responsibility in causing climate change is not equally distributed globally or within each society. The majority of historic and current greenhouse gas emissions have been or are generated by the most developed countries and these make up a little more than 20 percent of the world's population. In an increasingly global market economy, a large part of the emissions that occur in the underdeveloped world also stem from satisfying the needs and wishes of the first world.

Thus, the causes and consequences of climate change are at all levels closely associated with the growing disequilibrium in human development and the resulting moral, political, and ideological dilemmas. The Kyoto Protocol recognizes distinct levels of responsibility in the generation of climate change and in the distribution of responsibilities in policy responses. The poor societies are also the most vulnerable to climate alterations, and at the same time a major share of their potential development rests on natural resource extraction activities (energy, minerals, forestry, agriculture, and so on) that are among those that most affect the climate balance. In addition, these societies have fewer resources available to prevent or mitigate the consequences of climate change and are the most vulnerable to the indirect impacts resulting from the alteration of ecosystems, economic breakdowns, the appearance of new threats to health, or the degradation of vital resources such as water and soil.

On the other side of the scale, the most developed societies, with a little more than a fifth of the world's population, account for the highest rates of consumption per capita of fossil fuels and greenhouse gas emissions. In fact, the North–South inequality is larger in terms of resource consumption and the distribution of the environmental load than in terms of income or any of the other common economic indices used to compare levels of wealth and poverty across human societies. Moreover, climate change may sharpen existing inequalities and poses moral dilemmas that are difficult to resolve within the context of the current economic crisis and international geopolitical configuration. The painstaking processes of international negotiation to achieve a post-Kyoto protocol that includes concrete commitments from the emerging economies that have until now been excluded from concrete reduction commitments—like India, China, Mexico, or Brazil—is evidence of the difficulties in resolving these dilemmas.

Action in a Labyrinth

The fact that individual contributions to climate change are partial, and often diffuse, makes it difficult for us to recognize ourselves as both causes of the problem and as agents for its solution. The ethical and social labyrinth dilutes individual responsibility and impedes a realization and adequate valuation of the different roles that we play in relation to the problem (as consumers, citizens, professionals, and so on). To live and act in the risk society, suggests Beck (1998), became something Kafkaesque in *strictu sensu* of the word, if the adjective 'Kafkaesque' can be held to describe the absurd situations that afflict an individual in a totalitarian world, inscrutable and rambling for him as well as impossible to be designated by any other notion.

This phenomenon is enhanced by the difficulty that people face in identifying their concrete contribution to climate change as well as the repercussions that it could have on their life or the lives of those in their immediate social setting. For example, most people have a distorted vision of the current 'energetic model' and are rarely capable of evaluating their contribution to it. They do not know where the energy they consume comes from or how it is produced, to a large extent they do not know how and how much they consume, and they do not identify the consequences that their energy-related behavior has on the environment and the climate. Distortions introduced by the market must be added to this situation, since energy prices tend not to be a reliable indicator of either the economic or environmental production costs. In this context, it is not easy to plan changes for something that is unknown and for which we have a partial, incomplete, and distorted experience and view.

The Invisible Multicausality

Developed societies—and increasingly also those that are not but that aspire to development imitating the model of the first world —base their functioning on the consumption of large quantities of energy obtained primarily from fossil sources. This energetic voracity means that practically all of our actions— including when we sleep—generate greenhouse gas emissions. In some cases the relationship between our everyday behavior and greenhouse gas emissions are evident (as in the car emissions that we can 'see'), but in the majority of cases are 'invisible' or pass unnoticed given the complexity of the system of production and consumption of products and services (e.g., greenhouse gas emissions linked to meat consumption). The invisibility of a large part of the processes that connect us to the problem contributes to personal inaction.

Incoherence Between Messages and Policy Responses

The warnings over the existence of climate change and of the urgency of moving to action in the personal sphere are many times inconsistent with public policies and the optimistic messages associated with consumer culture projected through advertising as well as from the other ways of shaping lifestyles. This contradiction contributes to buffer the social perception of the threat, and acts as a disincentive for the adoption of alternative behaviors that are difficult to realize in practice without structures and conditions that facilitate them (e.g., the use of public transportation, consumption of energy from alternative sources, etc.). People tend to perceive the slowness in the development of policy responses to climate change as a sign that the seriousness of the threat is not so great, that urgent actions are not required, and that there is still a large margin of time to act.

The 'Cult of Progress' and Anthropological Optimism

Our limited memory of the past and the linear vision of history that locates 'modern civilization' at the zenith of human evolution are the basis for a 'cult of progress' resting on the optimistic myth that the future will always and necessarily be better than the present. This myth of the Enlightenment assumes that our species has known how to overcome other critical moments in the past despite the fact that human history is littered with civilizations that have collapsed and others that have experienced decadence and notable slides backward (Diamond 2005; Berman 2007; Steiner 1971). Many of these cultures collapsed due to ecological problems, despite their capacity to create extremely sophisticated civilizations in their own time and space. This myth is also fed by the confidence that science, technology, and social creativity, as tools of advanced societies, will end up finding 'a solution' that does not require substantial changes in the established model.

Psychosocial and Communication Barriers

The third type of conditioning factors for social representation of climate change are related to the cognitive and psychosocial processes that mediate, reinterpret, and integrate information within the environmental culture of the population in order to construct the social representation of climate change. It is important to take them into account for at least two reasons:

1. to avoid falling into the politically and pedagogically naive thinking that it is sufficient to transfer scientific information about climate change to society—a conventional type of 'scientific literacy'—in order for the problem to be understood and assumed; and
2. to help orient and define, from the viewpoint of the social representation of climate change, the programs, actions, and educational initiatives for communication and participation that can train and mobilize the population with respect to the climate threat.

In the following discussion we will briefly describe some of the beliefs, conceptions and prejudices about climate change that are most extensive in the popular culture of the populations of developed societies according to our own research experience (Meira 2001, 2006, 2009; Meira and Arto-Blanco 2008), as well as that collected by other authors (for example, Bord et al. 1998; Ungar 2000; Zehr 2000; Heras 2003; Moser and Dilling 2004; FUTERRA 2005; Abbasi 2006).

Isn't a 'More Temperate' Climate Better?

The difficulty people face in perceiving how they will be directly affected by climate change can feed into the idea that benefits can be expected from a 'moderate' increase in temperature; after all, the IPCC's evaluations talk of increases in 'only' tenths of a degree per decade. It is worth remembering that the 'temperate climates' are valued as positive and desirable within the contemporary culture of leisure and wellbeing. The 'Mediterranean climate' is a cultural standard for 'good weather' and is strongly associated with the development of the tourism industry. In colder latitudes and for social groups whose activities are currently limited by thermic variables, the expectation of an 'increase in temperature' can be valued as a positive and desirable change (for example, for the productivity and enologic quality of certain grape varieties in Central and Northern Europe, or for a boost to 'sun and beach' tourism on Europe's Atlantic coastline).

If we daily experience and respond to temperature changes of various degrees between the minimums and maximums without substantially altering our lives, how could a decimal increase in average temperature of the planet over a much longer time period be considered a threat?

When Our Senses 'Betray' Us

Our sensory apparatus is prepared to capture the evolution of atmospheric weather and to respond adaptively to daily, or at most seasonal, temperature variations, but it is not designed to capture and register the subtle variations in temperature that are quantified in tenths of a degree per decade over large periods of time. The 'warming' is, in physical terms, imperceptible, and an acceptance of its existence becomes, literally, and in the first instance, a 'question of faith' and of trust in science. Given our need to give meaning to reality, within popular culture meteorological phenomenon that are part of the typical irregularity of the weather are easily interpreted as proof or counter-proof for climate change. Thus a Siberian cold front in a zone especially sensitive to climate change such as the south of Europe tends to be interpreted as proof against climate warming, while equally a heat wave in the same region will be interpreted as evidence of climate change even though it is a meteorologically 'normal' occurrence at these latitudes.

COGNITIVE BARRIERS LINKED TO THE PROCESSING OF INFORMATION

If the senses show little capacity to capture the physical signs of climate change, our capacity to select, process, and interpret information is also limited by the principles of economics and simplification in daily life. As a function of these cognitive limitations, the complexity and scale of climate change, in addition to the dimension of uncertainty, establish an important difficulty for its perception and representation by the population. As we have discussed, scientific information when transferred to popular culture, is processed using epistemological patterns that are different from those that govern its development in the scientific domain.

The specialized literature identifies some of the sociocognitive patterns within the population that, from this perspective, distort the representation of climate change and its potential threat. These are as follows:

- The tendency to perceive the atmosphere as a vast and 'empty' space, capable of absorbing everything and that has remained unaltered and unalterable throughout time. This common 'belief' contrasts with the scientific finding that it is a fragile system made up of multiple layers whose total width is very small in relation to the volume of the Earth, and that has changed in composition and dynamics during the natural history of the planet.

- The confusion between atmospheric weather and climate. The frequent confusion between weather and climate also hinders an adequate representation of climate change within popular culture.

- The tendency to think that warming is a linear process and trust that the changes will follow a gradual evolution in time that will thus permit us to progressively adopt adaptation and response strategies. This belief ignores that, in the temporal scale that governs climate evolution, the current warming has been extremely rapid and can cause sudden and unpredictable changes whose occurrence may be catastrophic if certain thresholds are passed.

- Humans tend to emphasize the importance of extreme environmental phenomena but we have difficulty understanding gradual and progressive changes in the environment. In fact, we tend to overvalue extreme meteorological phenomena, and associate them with climate change when they may or may not be related. While on the contrary, we tend to fail to appreciate or tend to undervalue the subtle but relevant changes that are occurring in, for example, the distribution of the large terrestrial biomes due to small increases in temperature.

- The perception that given the magnitude of the problem individual action is irrelevant. This sensation of being "overwhelmed" (Giddens 1993; Meira 2001) constitutes one of the main psychosocial barriers that hinders awareness of climate change and responsible action. That is, even when the problem is perceived, the perception can be that an individual response is irrelevant and useless given its scale.

- The social representation of climate change is being constructed reusing ideas, beliefs, and concepts that were constructed for the environmental crisis in general and for other specific environmental problems and makes generalizations and establishes connections and relationships that have little or nothing to do with climate change science. The difficulty that most lay people have in understanding climate change from a scientific viewpoint is compensated for by activating sociocognitive processes that permit the reuse of ideas and representations that have already been internalized and shared for other similar issues. The most emblematic example in the case of climate change is the extensive belief that there is a causal relation between the 'hole' in the ozone layer and climate change. This belief is so common that we could speak of a 'cultural universal' within advanced societies. It is also a paradigmatic example of how information and concepts generated in the scientific field are integrated and reprocessed in popular culture. The fact that citizens use socially established knowledge, whether it is scientifically valid or not, to make climate change intelligible can constitute an important obstacle for the policy responses since from the perspective of education and communication it is easier to create a new representation than to modify one that is already established.

The Perceived Cost of Change

Since the problem is closely linked to the styles of life that are enjoyed or desired by the majority of the population in advanced societies, the personal cost of adopting important changes is perceived as very high (in economic terms to renounce levels of wellbeing achieved, the loss of security, and so on). Under the tension generated by the contradiction between awareness of the problem and personal inaction, people tend to transfer responsibility to other actors for generating and thus also for finding solutions or alternatives to the problem (toward industries, governments, pressure groups, international organizations, politicians, and so on). Through this external attribution, individual responsibility is diluted in collective responsibility and suffers the paradoxes of the "tragedy of the commons" (Hardin 1968). That is, when personal action implies the sacrifice of certain levels of personal wellbeing—whether these are objective or subjective—for the benefit of others, individual inaction is explained, justified, and legitimated by collective inaction, whether this inaction is by others or is in the perceived inefficacy of institutional responses.

Climate Change is Very Low in the Hierarchy of Needs

Climate change appears in a scenario of multiple problems that present themselves to the individual as local or global threats. These span the range from international terrorism to the financial crisis and North–South inequality at the global scale; or the threatened collapse of the 'welfare state,' unemployment, and insecurity, as the local and everyday experience of the inhabitants of the developed world. This state of permanent crisis means that a sensation of insecurity and threat become 'normal,' making it difficult for people to make decisions about which threat requires priority attention. Climate change does not sit high in the hierarchy of concerns that is logically more centered on those problems that directly affect the here and now, and is more closely related to the satisfaction of immediate basic needs. Such a scale of priorities favors the tendency to avoid and delay those problems that are perceived as more abstract, distant, and less concrete.

CLOSING REMARKS: EDUCATING AND COMMUNICATING ACCORDING TO SOCIAL REPRESENTATIONS OF CLIMATE CHANGE

We want now to be especially concise. It is necessary to research more in depth how social representations of global environmental problems are constructed in different social collectivities and biogeographical contexts, and not just with groups of students or in school contexts. A better understanding of how 'popular culture' interprets, gives meaning to, and elaborates implicit theories and beliefs, and how these are used to guide action can contribute to improving communication and educational processes for these problems. This obliges us to go beyond 'transmission' approaches as well as those of classic constructivist research centered in identifying and correcting the errors in peoples' 'scientific' ideas. Rather, sociopedagogical and psychosocial perspectives should be integrated and promoted.

From this point of view, research should also be directed to the comparative analysis of the social representations of climate change in different societies and cultures. The majority, not to imprudently say all, of the available studies have been carried out for Western populations. Understanding how the 'world is constructed' and especially the 'contemporary world' from other ethnic, religious, cultural, and social perspectives can help preserve and enhance human diversity in the face of the dangers of cultural homogenization and the standardization of social representations in the media.

It is also necessary to review the slogan 'think globally, act locally' as a guide to educational and environmental action. Global environmental problems oblige us to think and act simultaneously in two dimensions. Revision of this principle in educational practice offers many possibilities, but we think it is necessary to begin by identifying in each concrete social context—at the local and regional level—the existing connections between everyday individual and communitarian practices and the causes and consequences of climate change. That is, illuminate in social experience what is hidden in the impenetrable complexity of the new global reality. The analysis and questioning, for example, of the social and environmental genealogies of consumer products, which are almost always hidden, distorted, or ignored, can contribute to this task.

It is important to introduce debate—understood as exchange, critical contrast, and negotiation of meanings—on global problems in the spheres of everyday life. Social interaction is fundamental in the adjustment and readjustment of the social representations of reality, but global environmental problems rarely become 'material' for symbolic exchange with others, except in school spaces, and not always then. It is urgent to avoid citizens operating like 'dead end alleys' with the information, whether scientific or of other kinds, that they receive. This debate (in neighborhood, cultural, professional, or consumer associations, in recreational places, in cultural spaces) should not only address the social representation of climate change, but also possible solutions—in the international and individual spheres—and connect both of these. It is important to develop the 'belief' that this is not an exclusively scientific matter and that the solutions should also emerge from the collective construction of new economic, social, cultural, and ethical patterns. It is important to create educational situations where the subject-observer or information recipient becomes a subject-interpreter and subject-social actor.

Facing these challenges, as environmental educators and communicators we need to review our representations of climate change (information, knowledge, interpretations, experiences, implicit theories, beliefs, values, dispositions to action, and so on) in order to clarify and orient our practices. We should also be conscious that the target audiences for our work already have their own ideas about climate change and how to act or not in relation to it. People are not automatically transformed when they receive new information or information that is closer to scientific truth. In the everyday representation of the world there are variables and processes (cultural, emotional, experiential, contextual, social) that are important to elucidate and should become part of our educational 'subject' and 'content.'

BETWEEN COLLAPSE AND MITIGATION: TWO SCENARIOS

In few environmental problems does the future play such a central role as in climate change. The main work of the IPCC has been focused precisely on drawing up the most probable future scenarios with a view to fine tuning the mitigation and adaptation measures that will be needed to avoid the worst of them (those that are close to an average global warming of +6°C), and to hold things at the minimum that can be expected, now considered unavoidable, due to the level of greenhouse gas emissions already recorded: global warming of +2°C. According to the most recent IPCC report (2007), the future will fall somewhere between these two extremes. Therefore, the future, to a certain extent has already been written. Reaching the +6°C threshold towards the end of this century will mean the collapse of human civilization as we know it; limiting warming to +2°C will evidence a radical change in the way in which we universally go about satisfying our most basic needs.

Collapse Scenario

The system of production, distribution and consumption of resources continues, mainly in the energy field, and so allows the accelerated destruction and degradation of the natural processes responsible for fixing atmospheric carbon, principally the vegetal biomass and oceans.

Most inhabitants of countries in the West are conscious of the threat posed by climate change, but their resistance to changing established lifestyles stays very high, especially on issues like mobility that are directly relevant to the way they conceptualize their standard of living. The system itself sends out contradictory messages: on the one hand, there is an increase in the amount of information available on the climatic threat and the gravity of the situation; on the other, conspicuous consumption of products and services is encouraged to ensure that the economic system, threatened by the financial crisis, does not collapse.

Messages demanding lifestyle change face the counter-brainwashing effect of the constant dribble of news on technological breakthroughs that claim to have solved the riddle of how to mitigate and reverse climate change; messages that feed the optimism of a society still faithful in its belief in the progress inherent in modernity and the inevitability of a better future. Improvements in the efficiency of some technologies—cars or electrical appliances—is offset by a psychological 'rebound' effect: by using the latest generation of combustion engines, or hybrid cars, people can get better mileage, but travel further and buy more cars, so aggravating the problem.

The consequences are increasingly grim. Inhabitants of regions most vulnerable to climate change attempt to migrate to other areas seen as safer, fleeing desertification, water shortage or contamination, lack of food supply, flooding of coastal areas due to rising sea levels, extreme climatic phenomena, or diseases that climate change has helped aggravate or spread to previously uninfected places. Internal conflicts in the least developed countries are increasingly violent and North–South tensions heighten, especially in the face of the avalanche of climatic refugees.

From an environmental education and communication standpoint, the 'boiling frog' effect kicks in, people not realizing or underestimating what is happening to them. Contrarians still cast doubts and uncertainties on the science of climate change and find a ready ear. Blind faith in the market economy and the alliance between science and technology feeds the hope that 'a' solution will appear, one that will not substantially change a lifestyle that affords so much wellbeing, at least to some people. Instruction and education on climate change still focuses on making society scientifically literate so that they can understand the problem, setting aside its political, economic, and moral dimensions.

Mitigation Scenario: A Lesser-Evil Utopia

Avoiding the most catastrophic scenario has required a drastic change in the carbon diet that once nourished the global economy as we try to achieve an 80 percent to 90 percent drop in emissions by 2100. This has meant a substantial lifestyle modification in the most advanced societies, a substantial change in mobility patterns, urban development, production and consumption, services development and, fundamentally, of the way in which wellbeing is understood and achieved.

In the least-developed societies, a new approach to development, all about meeting unsatisfied basic needs without degrading the climate, has been taken. This has gone hand in glove with a considerable first-world reduction in greenhouse gas emissions so as to achieve an agreed global emissions balance.

Lifestyle change has been radical and has involved cultural mutation on an unprecedented scale. In the advanced countries, the core changes have been in mobility patterns and reducing conspicuous consumption. This has needed new understandings of wellbeing and status. In the consumption field, a "personal carbon credit" system operates (Fleming 2006; Starkey and Anderson 2005). Everyone is assigned an annual emissions quota with which to satisfy their total energy needs. If they exceed their quota, they have to buy from other citizens who have managed to organize their lives to emit less than their assigned quota and can sell the surplus carbon.

To achieve cultural mutation, coercive educational and communicative instruments have been used in tandem with structural change (in areas like transportation, urbanism, production, distribution, and consumption) and solar, geothermal, and wind energy development.

NOTES

1. Translation into English by David Tecklin, david.tecklin@gmail.com, and Ian Gardner, iangardner@prodigy.net.mx.
2. Deep ecology is a radical tradition of environmentalism that accords value to every kind of life including nonhuman life. It has been a well-recognized basis of the green movement in the developed world. One outstanding representative of this current of thought is Arne Næss, who proclaims that all the species and the living environment as a whole have the same rights to live and flourish as humankind. On the other side, social ecology is one of the most influential leftist currents of environmentalism. It is located within anarchism. One conspicuous representative of this tradition is Murray Bookchin, who holds that present environmental degradation and its problems are rooted in hegemonic political and social systems. See Devall and Sessions (1985) and Bookchin (1982).
3. "Scientific literacy is the capacity to use scientific knowledge, to identify questions and to draw evidence-based conclusions in order to understand and help make decisions about the natural world and the changes made to it through human activity" (OECD 2003, 131).

REFERENCES

Abbasi, D. R. 2006. *Americans and climate change: Closing the gap between science and action.* New Haven, CT: Yale School of Forestry and Environmental Studies.

Aldrich, B. and J. Kwong. 1999. *Educación medioambiental. (Environmental education).* Madrid: Círculo de Empresarios.

Beck, U. 1998. *Políticas ecológicas en la edad del riesgo. (Ecological politics in the age of risk).* Barcelona: El Roure.

Berman, M. 2007. *Dark ages America. The final phase of empire.* New York: Norton.

Bookchin, M. 1982. *Ecology of freedom: The emergence and dissolution of hierarchy.* Palo Alto, CA: Cheshire Books.

Bord, R. J., A. Fisher, and R. E. O'Connor. 1998. Public perceptions of global warming: United States and international perspectives. *Climate Research* 11: 75–84.

Capra, F. 2003. *A teia da vida. (The web of life).* São Paulo: Cultrix.

Caride, J. A. and P. A. Meira. 2001. *Educación ambiental y desarrollo humano (Environmental education and human development).* Barcelona: Ariel.

Devall, B. and G. Sessions. 1985. *Deep ecology. Living as if nature mattered.* Layton, UT: Peregrine Smith Books.

Diamond, J. 2005. *Collapse: How societies choose to fail or succeed.* New York: Viking-Penguin

Flannery, T. 2006. *We are the weather makers: The story of global warming.* Boston: Atlantic Monthly Press.

Fleming, D. 2006. *Energy and the common purpose. Descending the energy staircase with tradable energy quotas.* London: The Lean Economy Connection.

Foladori, G. 2000. El pensamiento ambientalista. *(Environmentalist thinking). Tópicos de Educación Ambiental* 2 (5): 21–38.

FUTERRA. 2005. *The rules of the game. Principles of climate change communications.* London: DEFRA—Climate Change Communications Working Group.

Giddens, A. 1993. *Consecuencias de la modernidad. (The consequences of modernity).* Madrid: Alianza Editorial.

González-Gaudiano, E. 2007. *Educación Ambiental: Trayectorias, rasgos y escenarios. (Environmental education: Paths, features and settings).* México: Plaza y Valdés-UANL/Instituto de Investigaciones Sociales.

Hardin, G. 1968. The tragedy of commons. *Science* 162: 1243–48.

Heras, F. 2003. Conocer y actuar frente al cambio climático: Obstáculos y vías para avanzar. *(Knowing and acting in front of climate change: Obstacles and ways to go forward). Carpeta Informativa del CENEAM,* Dec., 74–82.

IPCC. 2007. *Climate change 2007: Synthesis Report. Contribution of Working Groups I, II and III to the Fourth Assessment Report of the Intergovernmental Panel on Climate Change.* Geneva: IPCC.

Lomborg, B. 2001. *The skeptical environmentalist: Measuring the real state of the world.* Cambridge: Cambridge University Press.

Lyotard, J. F. 1984. *The postmodern condition: A report on knowledge.* Minneapolis: University of Minnesota Press.

Meira, P. A. 2001. Problemas ambientales globales y educación ambiental: Una aproximación desde las representaciones sociales del cambio climático. *(Global environmental problems and environmental education: An approach from social representations of climate change). IV Encuentro Internacional. Formación de Dinamizadores en Educación Ambiental: Investigación, Educación Ambiental y Escuela.* Medellín: Ministerio de Educación de Colombia.

Meira, P. A. 2006. A loita pola representación social do cambio climático: Unha reflexión para educadores. *(A struggle for the social representation of climate change: An insight for educators).* In *O cambio climático e Galiza,* ed. Soto, M. and X. Veiras, 53–61. Santiago de Compostela: ADEGA.

Meira, P. A. and M. Arto-Blanco. 2008. De la conciencia a la acción. La representación del cambio climático en la sociedad española. *(From the awareness to action. The representation of climate change in Spanish society). Seguridad y medio ambiente* 108: 31–47.

Meira, P. A. 2009. *Comunicar el Cambio Climático. Escenario social y líneas de acción. (Communicating about climate change. Social settings and ways of action).* Madrid: Ministerio de Medio Ambiente y Medio Rural y Marino-Organismo de Parques Naturales.

Moser, S. C. and L. Dilling. 2004. Making climate hot: Communicating the urgency and challenge of global climate change. *Environment* 46 (10): 32–46.

OECD. 2003. *PISA 200. Assessment framework: Mathematics, reading, science and problem solving knowledge and skills.* Paris: OECD.

Orr, D. W. 1994. *Earth in mind: On education, environment, and the human prospect.* Covelo, CA: Island Press.

Starkey, R. and K. Anderson. 2005. Domestic tradable quotas: A policy instrument for reducing greenhouse gas emissions from energy use. *Technical Report 39,* Tyndall Centre for Climate Change Research. http://www.tyndall.ac.uk/research/theme2/final_reports/t3_22.pdf (accessed April 15, 2009).

Steiner, G. 1971. *In Bluebeard's Castle: Some notes towards the redefinition of culture.* London: Faber & Faber.

Sterling, S. 1996. Developing strategy. In *Education for sustainability*, ed. Huckle, J. and S. Sterling, 197–211. *London*: Earthscan.

Sterling, S. 2001. *Sustainable education: Re-visioning learning and change*. Bristol: Arrowsmith.

Ungar, S. 2000. Knowledge, ignorance and the popular culture: Climate change versus the ozone hole. *Public Understanding of Science* 9: 297–312.

Zehr, S. 2000. Public representations of scientific uncertainty about global climate change. *Public Understanding of Science* 9: 85–103.

2 'Go, Go, Go, Said the Bird'
Sustainability-related Education in Interesting Times

David Selby

AN UNSUSTAINABLE RENDITION OF SUSTAINABILITY

At a time when climate change seemed a fairly remote and, at worst, slowly and incrementally unfolding prospect, the World Commission on Environment and Development issued its report, *Our Common Future*, defining sustainable development as "development that meets the needs of the present without compromising the ability of future generations to meet their needs" (WCED 1987, 43). This definition, rehearsed catechism-like in the pronouncements of sustainability educators through the last decade of one century and the first decade of another, becomes problematic given what is now understood about climate change. First, we better appreciate that the impacts of human-induced climate change are "seriously backloaded," a "substantially deferred phenomenon" (Gardiner 2008, 31–32). Hence, the effects of global heating[1] that we are currently experiencing are primarily the outcome of CO_2 emissions from some time in the past (ibid.). Second, carbon build-up in the atmosphere is an inexorable phenomenon that is not easy, and may be impossible, to moderate or reverse within a crisis-avoiding timescale, unfolding as it does to its own timetable (ibid.). The first point calls into question our agency in contributing to the wellbeing of at least more immediate future generations (whose carbon fate may already be more or less sealed) and also the possibility of our ever being held specifically and primarily accountable by future generations for any climate change effects. The latter point also feeds into a sense of our helplessness about promoting 'development' of a kind that we can be assured will be of benign near-time intergenerational effect. The impact of efforts to enable future generations to meet their own needs appears indeterminable. The complexity of climate change thus renders elusive the connection of cause and effect and so provides fertile ground for a 'business as usual' disposition to prevail and for self-deception to flourish. "Climate change may be a perfect moral storm," writes Stephen Gardiner (ibid. 37):

> . . . its complexity may turn out to be *perfectly convenient* for us, the current generation, and indeed each successor generation as it comes to

occupy our position. For one thing, it provides each generation with a cover under which it can seem to be taking the issue seriously—by negotiating weak and largely substanceless global accords, for example, and then heralding them as great achievements—when really it is simply exploiting its temporal position. . . . By avoiding overtly selfish behavior, an earlier generation can take advantage of the future without the unpleasantness of admitting it—either to others, or, perhaps more importantly, to itself [italics in original].

Much of the discourse of sustainable development, and of its educational outcropping, education for sustainable development, has become steeped in 'business as usual' assumptions by explicitly or implicitly (i.e., through definitional inference or evasion) interpreting development as connoting continued economic growth. Sustainable development, a term easily susceptible to what Lucie Sauvé (1999, 19) describes as "semantic inflation," has largely been captured by neoliberal market thinking and a corporatist agenda that projects economic growth as an immutable necessity rather than a pathway of choice. As such, it has become part of the problem rather than part of the solution, for if we accept the finiteness of the planet—that the planet is not an inexhaustible cornucopia—and if we interpret 'sustainable development' as 'sustainable growth,' then the label becomes oxymoronic, a contradiction in terms, a "self-contained *non-sequitur* between noun and modifier" (Disinger 1990, 3). Albeit part of the problem, it does provide a palliative, a sense we are doing something while remaining deeply complicit in a growth paradigm that is destroying both ecosphere and ethnosphere.

Moacir Gadotti (2008, 25) adheres to the idea of education for sustainable development but recognizes that "without social mobilization against the current economic model" it "will not reach its goals." The field is at a forked path where it must choose between "competitive globalization" following the laws of the market and in which "peoples' interests are less important than the corporative interests of big transnational companies" or "planetization" informed by ethical values and spirituality (ibid. 23). Put another way, to avoid planetary catastrophe, a predatory view of development has to be replaced "by conceiving of it more anthropologically and holistically and less economically" (ibid. 31). Richard Kahn (2008, 9) calls for the ecopedagogical uncloaking of sustainable development to divest it of "the sort of ambiguity" that gives scope to the neoliberal impulse to "autocratically modernize the world despite the well-known consequential socio-cultural and ecological costs." For James Lovelock, sustainable development is a quixotic idea of archaic uselessness in tackling climate change. "Two hundred years ago, when change was slow or non-existent," he writes (2006, 3), "we might have had time to establish sustainable development, or even have continued for a while with business as usual, but now is much too late; the damage has already been done. To expect sustainable development or a trust in business as usual to be viable policies is like expecting a lung cancer victim to be cured by stopping smoking."

As Vandana Shiva (2008, 14, 16) sees it, development "cannot be defined by the colonizer," yet the notion has been precisely appropriated by marketplace globalization, what she terms "eco-imperialism," a mechanistic paradigm "based on industrial technologies and economies that assume limitless growth." The eco-imperialist response to climate change, at first one of refutation, and now, in the face of irrefutable evidence, morphing into one of busy engagement, has been to "offer the disease as the cure" (ibid. 2). In the guise of 'sustainable development,' the favored pseudosolutions to climate change have been further growth-oriented global marketplace developments, such as "ethically perverse" and "ecologically perverse" carbon trading, as well as technological innovation, including efforts to resurrect the nuclear power industry (ibid. 5, 19, 20). "Market fundamentalists are so committed to defending and enlarging markets," Shiva (ibid. 20) concludes, "that even when it is clear that climate change demands a radical change in our ways of thinking and our patterns of production and consumption, their preoccupation is with protecting market mechanisms, not Gaia's ecological process."

The same marketplace and technological orientations and avoidance of radical, self-questioning ways of thinking about the looming climate change crisis are manifest in policies, proposals, and programs for education for sustainable development in schools and universities. The Department for Children, Schools and Families for England and Wales offers a *Sustainable Schools National Framework* (DCSF 2006) comprised of eight 'doorways' for initiating sustainability-related school activities. They are: Food and Drink (healthy, local, environmentally friendly, and socially responsible food); Energy and Water (energy efficiency, renewable energy, sustainable water management); Travel and Traffic (reduction in vehicle use, alternative forms of transport); Purchasing and Waste (procurement to ethical standards); Buildings and Grounds (sustainable-building design; environmental management system); Inclusion and Participation (schools as models of social inclusion); Local Wellbeing (schools as models of good corporate citizenship); Global Dimension (sustainable development; schools as models of good global citizenship). Worthy elements in the response to climate change that the doorways for the most part are, they more or less adhere to a 'business as usual' agenda built upon the idea that the path to sustainability lies with a combination of better management, more technological efficiency, and responsible citizenship (usually inferring citizenship that does not overly rock the boat of 'business as usual' production and consumption). The human societal and individual condition as source and future victim of climate crisis is barely addressed by the sustainable school as so defined. In a similar vein, the World Wide Fund for Nature (WWF-UK) activity pack for teachers, *Climate Chaos: Learning for Sustainability*, declares that climate change "is considered to be the most serious environmental challenge facing our planet" and that "our way of life is threatened" (WWF-UK n.d., 2) before going on to offer, through classroom activities, scientific understandings of climate change, understandings of technological solutions, and reflections on benign consumerism, but without ever directly exploring why and how "our way of life," now threatened, is itself fundamentally culpable in fuelling the looming crisis. There is a tendency throughout the present genre of climate change educational materials to characterize the global heating crisis in terms of overtly presenting cause, that is, as a CO^2 problem curable within largely present terms of reference, rather than a problem arising out of the crisis of the human condition, a crisis arising from a disconnect from the web of life, especially among privileged populations, and, hence, a crisis of exploitation and violence coupled with denial. Alastair McIntosh (2008), reflecting on responses to the prospect of rampant climate change, observes:

> The question of whether technology, politics and economic muscle can sort out the problem is the small question. The big question is about sorting out the human condition. It is the question of how we can deepen our humanity to cope with possible waves of war, famine, disease and refugees without such outer wounds festering to inner destruction. (191)

Linking to the uncritical or tacit embrace of economic growth fed by globalization and consumerism, much of the field of education for sustainable development further offers disease as cure by representing nature as resource. There are relatively few voices among its proponents arguing that sustainability means a return to right relationship with nature or that nature is of formative importance for a centered humanity. An instrumentalist valuing of the other-than-human, that is a valuing according to usefulness in meeting perceived human imperatives, prevails, with a correlative denial of the intrinsic value of the natural world and of human embeddedness in nature (Selby 2007a, 258–61). In consequence, the field tends to turn in on its essentially anthropocentric core, the predominantly technical fix approach to climate change and other socioenvironmental threats prevalent among sustainable development advocates at large finding a ready echo in the primacy accorded to 'sustainability literacy' and 'sustainability skills and competencies for the workplace' on the part of many sustainable development educators and educational opinion formers (Selby 2006, 360–61). Through such "feeble offerings" (Lovelock 2006, 147), there is an urgency to align with the marketplace skills imperative but an overriding axiological avoidance. Concern for values, for the ethics of the human condition, tends to be pretty much left on the rhetorical shelf. And, yet, as is being suggested here, it is precisely a sleepwalked attachment to a distorted value system that is fuelling rampant and runaway climate change. "If we do not think that our own actions are open to moral assessment, or that various interests (our own, those of our kin and country, those of distant people, future people, animals, and nature) matter," writes Gardiner (2008, 25), "then it is hard to see why climate change (or much else) poses a problem. But once we see this, then we appear to need some account of moral responsibility, morally important interests, and what to do about both."

DENIAL AND COGNITIVE DISSONANCE IN RESPONSES TO CLIMATE CHANGE

Why the reluctance to wholeheartedly and comprehensively engage with the climate change threat?

Writing amid the so-called 'credit crunch' of 2009, it is remarkable (but by no means surprising) how much the reverberations from a failed banking system threatening to undermine the financial prosperity of a global minority have eclipsed a 'climate crunch' threatening humanity's future existence on the planet. "This is a most dangerous state of affairs," writes Jess Worth (2009, 1). "It's like finding out that you've got cancer, but then delaying going to the doctor's for treatment for a few months because you want to repaint your house."

This is but a particularly striking example of the "eyes wide shut" syndrome (Hilmann et al. 2007, 72 et seq.) that characterizes much of the general response to climate change. From government, media, educational institutions, and the public often comes a presenting acceptance, often fulsome, of the severity of the looming global heating crisis coupled with an ill-preparedness to follow through in terms of confronting the deep personal change and societal transformation needed to have any chance of staving off the worst effects of climate change. As such, climate change response constitutes a form of furtive denial characterized by fully conscious or threshold-of-consciousness cognitive dissonance between perception of problem and identified and acted upon (or not acted upon) response. It is a denial rooted in "cultural intoxication," suggests Diarmuid O'Murchu (2004, 140), the addiction of the Western world to material acquisition that drowns out an alienation from the life and death realities of the world and so upholds an illusion of power and control (ibid. 139). "We live in a dark age," O'Murchu continues (ibid. 140),

> but, alas, nobody wishes to entertain that notion. We are unable to befriend the darkness because our addictiveness and compulsiveness keep us firmly rooted in *denial*. The whole thing is too painful to look at, so we choose to befriend our pathology rather than befriend its deeper truth [author's italics].

This cultural condition of prevailing cognitive dissonance has elsewhere been metaphorically depicted as "cognitive polyphasia," a 'polyphase' being an electrical device in which alternating currents coexist (McIntosh 2008, 88); also, as a state of "consensus trance reality" (Tart 1986), that is, a condition of accepting the status quo and our powerlessness to change it on account of a more or less ubiquitous quasi-hypnotic condition that "helps to keep the bogey man away" (McIntosh 2008, 99). Mayer Hillman et al. (2007, 85) refer to "a near-universal state of denial, close to collective amnesia, about the significance of climate change," while Joanna Macy and Molly Young Brown (1998, 26) remind us of the etymology of the word 'apathy,' the Greek *apatheia*, literally the inability or refusal to experience pain, in explaining both our addiction to what they call the "Industrial Growth Society" and our repression of concern over its planetary impacts (see, also Selby 2007a, 254). "Go, go, go, said the bird: human kind cannot bear very much reality," writes T. S. Eliot in *Four Quartets* (1963, 190).[2]

In the prescient words of Sandra Postel (1992, 4):

> Psychology as much as science will . . . determine the planet's fate, be-cause action depends on overcoming denial, among the most paralyz-ing of human responses. While it affects most of us in varying degrees, denial runs particularly deep among those with heavy stakes in the status quo. . . . This denial can be as dangerous to a society and the natural environment as an alcoholic's denial is to his or her family. Be-cause they fail to see the addiction as the principal threat to their own well-being, alcoholics often end up destroying their own lives. Rather than facing the truth, denial's victims choose slow suicide. In a similar way, by pursuing lifestyles and economic goals that ravage the envi-ronment, we sacrifice long-term health and well-being for immediate gratification—a trade-off that cannot yield a happy ending.

The uncritical adherence to the growth principle that marks much of the field of education for sustainable development and the correlative focus on managing 'natural resources' or 'natural capital' or 'ecosystem services' (Porritt 2005, vi) and on technical fixes to address climate change are themselves manifestations of denial and cognitive dissonance. There is rec-ognition of the probability of looming catastrophe but the remedies most frequently raised both in school and university cocoon the learner by avoid-ing addressing the roots of the climate crisis while oftentimes barely raising the possibility that the remedies presented might be complicitous expres-sions of the crisis. Quantum physicist David Bohm was fond of recounting the Sufi story of the man who lost the key to his house:

> He was found to be looking for it under a light. He looked and looked and couldn't find it. Finally someone asked where he had lost the key. He answered, "Well, I did in fact lose it over there." And when asked why he didn't look for it over there, he said, "Well, it's dark over there, but there is light here for me to look." (Bohm and Edwards 1991, 17)

Failure to engage with the disorientation of darkness deepens the tilt of an already inclined plane. Some climate change scientists meeting in Copenhagen in March 2009 attested to their reluctance to state publicly their growing conviction that global warming targets are not likely to be met in that "the subject was sensitive" and giving it an airing would become a "self-fulfilling prophecy" in that the knowledge would de-motivate people (Adam 2009, 14). But, to exempt from knowing is to compound denial and its effects, while taking away both agency and responsibility. "We must begin by winning the first battle against inertia and the fear of change," argues Al Gore (2007, 196). "And this means we have to understand exactly what we're up against."

EDUCATION FOR SUSTAINABLE CONTRACTION AND SUSTAINABLE MODERATION

While recognizing that there are alternative autopoetic (self-organizing, communitarian) renditions of 'development' as against the dominant allopoetic (externally imposed or neocolonial) understandings of the term (Shiva 2008, 13–14), I am proposing that we consider the idea of education for sustainable contraction as better capturing a more realistic educational response to the prospect of imminent runaway global heating, taking the learner out of and beyond a state of denial. Alternative renditions of 'development' are now too easily drowned within a sea of growth-speak. In proposing education for sustainable contraction as an appropriate educational response to climate change, I acknowledge my indebtedness to James Lovelock. "What we need," writes Lovelock (2006, 7), "is a sustainable retreat." Lauding the orderly quality of the Napoleonic retreat from Moscow in 1812 and exhorting a 1940 Dunkirk spirit in the face of runaway climate change, he adds:

> We need the people of the world to sense the real and present danger so that they will spontaneously mobilize and unstintingly bring about an orderly and sustainable withdrawal to a world where we try to live in harmony with Gaia. (150)

Writing elsewhere, I express preference for "the softer and more ecological concept of 'contraction,' a concept also devoid of militaristic connotations and tending to infer the systemic rather than the systematic, the organic rather than the lockstep" (Selby 2007a, 258).

If education for sustainable contraction plays its part in heroic efforts especially across the developed world to effect a transformation to less exploitative and environmentally harmonious life ways, and, if the contraction project is ultimately somewhat successful in mitigating global heating, the concept may eventually morph into the more steady state idea of education for sustainable moderation.

In the following sections, I discuss key and complementary aspects, as I see them, of education for sustainable contraction and education for sustainable moderation, all of which are absent from or at best at the margins of the field of education for sustainable development.

ALTERNATIVE CONCEPTIONS OF A 'GOOD LIFE'

In 1981 Duane Elgin published his important treatise, *Voluntary Simplicity: Toward a Way of Life that is Outwardly Simple and Inwardly Rich*, a work little referenced in the literature on education for sustainable development. The work lays out principal tenets of an approach to living simply in the world that, in contradistinction to the dominant Western way, combines frugal consumption, ecological awareness, and personal growth. They include:

- Evolving both material and spiritual aspects of life in harmony and balance (as against seeing material progress as the overriding goal)

- Emphasizing conservation, need-limited frugality, and living a satisfying life in cooperation with others (as against indulging in conspicuous consumption)

- Conceiving identity as a fluid process of living (rather than being defined through accumulation of material possessions and social position)

- Seeing the individual as individuated, unique yet inseparable from the whole (as against individualistic and separated)

- Seeing the universe as living organism, honoring its preciousness and integrity (as against as reservoir of material resource for exploitation)

- Emphasizing life-serving behavior, giving as much as possible, and asking only what is required in return (as against self-serving behavior bent on acquiring as much as possible)

- Emphasizing connectedness and community (as against autonomy and mobility). (Elgin 1981, 40)

For Elgin anything lost in the transition to voluntary simplicity is more than made up for through the quality of revitalized experience and not least through the cultivation and application of "conscious watchfulness," the ability to see the close-to-hand world through intimate eyes, and also by means of "self-reflective consciousness," "the ability to be simultaneously concentrated (with a precise and delicate attention to the details of life) and mindful (with a panoramic appreciation of the totality of life)" (ibid. 149–151).

A countercultural idea informing the more recent Slow Food Movement (Petrini and Padovani 2006; Petrini 2007) and wider Slow Movement (Honoré 2005), the idea of voluntary simplicity also finds ready echo in the proposals of Vandana Shiva for Earth Democracy based on an age of "soil not oil" (Shiva 2008). For Shiva, Earth Democracy is a localized communitarian democracy in which those making up the local community, working within a local "living economy" and, hence, most intimately attuned to the local ecosystem, have the "highest authority on decisions related to the environment and natural resources and to the sustenance and livelihoods of people" (Shiva 2005, 9–11). "The real solution must be to search for right living, for well-being, and for joy while simultaneously reducing consumption," she writes (44), and continues:

> Earth Democracy allows us to break free of the global supermarket of commodification and consumerism, which is destroying our food, our farms, our homes, our towns, and our planet. It allows us to re-imbed our eating and drinking, our moving and working, into our local ecosystems and local cultures, enriching our lives while lowering our consumption without impoverishing others. (Shiva 2008, 46)

As will be mooted later, this idea has important implications for the nature and focus of citizenship education in a climate-changing world.

Education for sustainable contraction/moderation in the face of the climate change threat would seek to give school-age and adult learners a lived understanding of alternative conceptions of a 'good life,' and not least through the revival of local craft learning, attendant apprenticeships, and the active integration of learning within community initiatives. The curriculum would follow the rhythm of the seasons emphasizing the nuances of seasonal shift. Learners would be engaged in the chain of food production as growers, preparers, and consumers. They would work and learn in and with community gardens, preventative health care centers, childcare centers, home redesign endeavors, and the renaissance and diversification of forms of culture, storytelleing and community narrative.

On the opposite side of the same coin, education for sustainable contraction/moderation would incorporate anticonsumerism education. Inspired by the unremitting efforts of antiracist and antisexist education to eradicate overt and insidious discrimination on grounds of race, ethnicity, and gender, I propose the Herculean task of uprooting the life-sapping contamination of rampant consumerism—"consumption beyond the level of dignified sufficiency" (McIntosh 2008, 180)—as a key strand of efforts to achieve sustainable contraction and moderation though education.

It has been a key feature of 'light green' environmentalism to promote 'green consumerism,' often by employing the "reduce, reuse, recycle" legend. The problem with consumer awareness strategies is that they fall short of a root-and-branch critique of consumerism as such and may even inadvertently buttress a consumerist ethic. The recycling bin in most classrooms is a case in point. Often identified as evidence of a school's commitment to sustainability, it conveys the subliminal message that consumerism if approached responsibly can be benign.

For Sue McGregor (2003a, b), consumerism is both source and manifestation of structural violence, not only exploiting the indentured slave, the woman and child sweatshop worker, and the natural environment, but also enslaving the consumer herself. "People behave as they do in a consumer society," she writes (2003a, 3), "because they are so indoctrinated into the logic of the market that they cannot 'see' anything wrong with what they are doing. Because they do not critically challenge the market ideology, and what it means to live in a consumer society, they actually contribute to their own oppression." Key to that oppression is the failure to fulfill their innate potential by permitting consuming to make and shape their identity. Consumerism, she avers, "is an acceptance of consumption as a way to self-development, self-realization and self-fulfilment. In a consumer society, an individual's identity is tied to what she or he consumes" (ibid. 2). Alastair McIntosh concurs. For him, consumerism "interrupts the very journey of life. It keeps us narcissistically at a child-like level of immaturity, seeking only the next fix" (2008, 176). Borrowing from Richard Rohr (2004), McIntosh (2008, 175) draws the distinction between *liminal* experience (threshold crossing ways of thinking and experiencing that expand consciousness) and *liminoid* experience (momentarily renewing experiences, with a 'feel good' effect, that merely reaffirm ego and capacity for denial) of which consumerism is an example par excellence.

Anticonsumerism education has, then, the twin goal of protecting the ecosphere and ethnosphere while liberating the individual from the thrall of consumerism for a journey of self-discovery and self-growth. As such it dovetails with a learning journey to experience alternative, seemingly simpler but essentially richer, modes of living and relating within community. "We are called upon to become *connoisseurs*," writes McIntosh (2008, 234). "That's how to make more out of less and be the richer for it. Far from being a recipe to kill joy, it's the only sustainable way to en-joy" [italics in original].

Education directed towards fostering alternative, nonconsumerist, ways of being and becoming can make a significant contribution to the retreat from a growth economy, the backcloth to which will need to be age-appropriate learning windows for considering and concretizing ideas for transition to a slow-growth or no-growth economy. "As someone who lives in a rich country I would prefer to live under the current system, but climate change means we don't have a choice," says Peter Victor, author of *Managing Without Growth: Slower by Design not Disaster* (2008). "We can either design a slower-growth economy over the next few decades, or we'll get there suddenly through environmental disaster" (Chakrabortty 2009).

INTIMACY WITH SELF AND NATURE

Education in the Western world has, in the past thirty years, come to be infected by the Galileo fallacy, the view that what can be measured is primary (Capra 1983, 39). Accountability and performance, it has become commonplace to hear, must be determined and assessed according to quantifiable data. The management and technical orientation of much of the field of education for sustainable development has, in the main, led its proponents to collude with the predominating marginalization of the intangible.

It was during the time of Galileo (1564–1642) that T. S. Eliot says a "disassociation of sensibility set in, from which we have never recovered" (Eliot 1921 cited in McIntosh 2008, 154). This "breaking up of the ability to feel and relate to life," in McIntosh's view, lies behind "the mindlessness that underlies anthropogenic climate change" (112). Divorced from the ground of being, what physicist David Bohm calls the "holomovement " or flow of "unbroken wholeness" informing reality (Selby 2007b, 167–68), many, especially in the hegemonic West, suffer the illusion of experiencing themselves as "isolated egos in this world" (Capra 1983, 29), cutting themselves off from "outer confirmation of (their) inner life" (Zohar 1990, 217) and surrendering to "the current endemic sense of alienation and meaningless" (Charlton 2008, 8). To fill the void, the liminal deficit, there is an opting for liminoid experience.

Following from such insights, it would seem evident that two significant and mutually enfolded aspects of education for sustainable contraction and moderation would be, first, a reengagement with dynamical interconnectedness as a form of spirituality, and, second, a recultivation of the poetic, of the metaphysical, of ecospiritual ways of knowing.

Inspired by the work of Gregory Bateson, Noel Charlton argues (2008, 160) that "the only social process that seems potentially to be able to override the consumerist, aggressive power-hungry ethic that is prevalent now, seems to be a psychological dynamic towards the sacred nature of the total ecology." Bateson defines the sacred as the "vast interconnected whole that is the totality of all the nested systems or minds of the living world" (ibid. 163). Throughout nature and existence, from simple organisms to what we have come to call Gaia, he held that there were scales of mind of which the individual human mind is a subsystem immanent within its own body but also receiving and transmitting knowing from and to the larger scales of culture and nature (ibid. 164). Looming climate change disaster according to a holistic worldview of this kind comes about through the illusion of disconnection and immunity from, and hence superiority over, other-than-human and larger-than-human mind.

Importantly, Bateson identifies art and other forms of aesthetic engagement as important pathways in achieving "the grace of relatedness" (ibid. 137). In this regard, it is significant how little space is given to reawakening a sense of beauty within the canon of education for sustainable development literature. In calling for the recultivation of the poetic, so resensitizing the learner both to the numinous and to her full panoply of potentials, I am calling for the recovery of emotional and eco-spiritual, but decidedly non-measurable, ways of knowing and ways of connecting to the world—such as attunement, awe, celebration, enchantment, intuition, reverence, wonder, and the oceanic sense of connectedness. In her lyrical book, *The Sense of Wonder* (1965, reprinted 1998), Rachel Carson offers her learning credo. "I sincerely believe," she affirms (1998, 56), " that for the child, and for the parent seeking to guide him, it is not half so important to *know* as to *feel"* (italics in original). Giving primacy to feeling responses during early years can better enable the marriage of sense and sensibility (Selby 2006, 301) as equal and complementary partners in mature adulthood. In so doing, we begin to address a fundamental problematic in how we approach the climate hazard; that is, the adequacy of rationality in "resolving issues as complex, subtle and multidimensional . . . as environmental concern, not least from the motives and values embedded in modern rationality *itself*," expressing, as it does, "certain aspirations towards the world, notably to classify, explain, predict, evaluate and, as far as modern rationality is concerned, increasingly to exploit it" (Bonnett 1999, 321). "If we learn, before it is too late," writes Charlton (2008, 161), "to make this move towards reverential relationship with the systemic and material world, we will be in a win-win situation. We will gain enormously in quality of life. We will cease to be a pathology within the systems of the living Earth."

NONVIOLENCE EDUCATION

The "disassociation of sensibility" is a form of violence—to self—and, in the playing out of the divided, disconnected, and alienated self, to other humans and the planet in the shape of direct physical violence and violence and exploitation embedded in hegemonic structures and systems. McIntosh (2008, 131) offers an insightful formula in this regard: hubris = pride $\rightarrow$ violence $\rightarrow$ ecocide. Hubris, the presumption of seeing "man as the measure of all things," he says (217), is both manifest in and a manifestation of the pride of the "ungrounded and inflated ego" (109), social violence and repression (227), and ecocide, the extension of direct and structural violence to nature. Drawing from his formula, the looming prospect of runaway climate change can be perceived as the most threatening consequence of a culturally embedded hubristic violence that has acquired global reach through the cultural imperialism of the marketplace, pushing holistic understandings of humanity's relation to nature to the countercultural and indigenous margins.

While sustainability-related education was early conceived of as encompassing issues of peace and social justice, its proponents have essentially mapped out the field along the development/environment interface (Selby 2006, 361). Education for sustainable development has, at best, only hesitantly taken onboard the insights, concerns, and conceptual frameworks of peace, social justice, and antidiscriminatory education, something I suggest should be remedied by an education looking to sustainable contraction and moderation. Confronted with the looming presence, and soon immediacy, of rampant climate change, it would seem a matter of both expedition and survival prudence to investigate social and environmental crises as crises with violence at their root and, contrapuntally, to build and hone understandings of the theory and practice of nonviolence and nonviolent security. With the huge population displacements that are anticipated with the onset of global heating, involving massive influxes of populations into receding northern and southern "belts of habitability" (Lynas 2007, 225), learning programs focused on conflict avoidance and resolution, unpacking negative and enemy images of the 'other,' and addressing the hegemonic 'centrisms' that have served to foment injustice and discrimination (Smith and Carson 1998; Plumwood 1993a) would seem to be correlatively essential.

There is incipient and, in many respects, already conspicuous social injustice deeply embedded in climate change impacts and responses. While countries in the metaphorical South of the planet are held to account for their financial indebtedness, there is so far no commensurate holding to account of countries of the metaphorical North for their ecological indebtedness. Calling for a proper sharing of the "global atmospheric commons" allied with due restitution by those with a history of ecological appropriation, Sunita Narain (2009) writes:

> The tragedy of the atmospheric commons has been the lack of rights to this global ecological space. As a result, industrialized countries have borrowed heavily and without control. They have emitted greenhouse gases far in excess of what the earth can withstand, 'free riding' on the planet's natural capital. Some have called this the ecological debt of the North, which stands as a counterweight to the financial debt of the South. (11)

For Narain responding to climate change involves "redistributing economic and ecological space" through the global adoption of "equal per capita entitlements to greenhouse gas emissions" (ibid.). For Shiva (2008, 3, 47) compensation for those not primarily responsible for global heating is essential and particularly so given that the impacts of climate change fall in a hugely disproportionate manner on nations and communities in the South.

In responding to global heating, a sustainability-related education of relevance cannot but place climate change justice at the core of its agenda. There are deep-seated and unavoidable issues of global justice and restitution involved, for all to address and in pursuance of which conscientization programs, especially for Northern populations as they expurgate their "affluenza" (Hamilton and Denniss 2005), will be vital. Given the ever-present tendency of the economically privileged to construe climate change as an issue calling for a scientific, technological and individual behavioral reform response, it seems vital that "we start putting a human face to the climate change that is beginning all around us" to help bring justice issues to the table (Narain 2009, 11). "We must," writes Narain, "see climate change in the faces of the millions who have lost their homes in the cyclones that ripped through Bangladesh and Burma this year; in the faces of those who have lost everything in floods caused by extreme rainfall events" (ibid). In this regard, Elizabeth Kolbert's narratives of lives damaged by climate change, *Field Notes from a Catastrophe* (2007), is path finding. It is a matter of social justice education, too, as well as a potential contribution to survival, to bring to the learning communities of the ecological debtor nations of the North the nonappropriated voice and experience of the South, where the dynamics of nonaccumulative sustainable living are still the day-to-day reality for many communities (Okolie 2003).

James Lovelock (2009, 12, 56) ponders the terrifying question of how those places in the farther north and south, "lifeboats for humanity" least affected by global heating, will respond to the clamorous multitudes wanting to get on board. "Will the privileged," asks McIntosh (2008, 236–37), "pull up the drawbridge to try and keep at bay the human consequences of their consumer profligacy?" Before such extremes are reached, there will inevitably be a surge of internal migration from coastal cities and areas made uninhabitable by drought and wildfire, as well as intense pressure for increased immigration and repatriation. A thoroughgoing anticipatory program of nonviolence, climate justice education, and education in humanitarianism, of global reach, is called for to help stave off the kind of tribal and mob responses coupled with the suspension of democracy that Lovelock anticipates (ibid. 61).

LIVING AND LEARNING AS DENIZEN

Alongside the global ethic that climate change peace and justice demand, a program of education for sustainable contraction and moderation also requires a complementary localization of focus. In the weaning off consumerism and (re)learning of intimacy with self and nature through "conscious watchfulness," as proposed earlier, the bioregional and deep ecological notions of inhabitation and reinhabitation of place have an important bearing; that is, being or becoming native to a place, internalizing its particular natural and associated cultural characteristics, and shaping needs and livelihoods according to the land (Traina and Darley-Hill 1995, 4), just as the land is modified over time by those needs and livelihoods (Cannovó 2008, 190–91). There are innumerable environmental theorists who have identified connection to place as a basis for environmental responsibility and ecological values. For Val Plumwood (1993b, 297) "deep and particularistic attachment to a place" is not only identity forming, but expresses itself in "very specific and local responsibilities of care." For Barry Lopez (1998,

132), place is vital for identity, for "a feel for the soil and the history." Place attachment can, thus, be a significant source for ethical responsiveness to climate change. In fostering localism and bioregionalism as responses to global heating, it may be that environmental education needs to return to its 'nature study' roots. Developing knowledge and appreciation of regional flora and fauna and of seasonal weather patterns has been proposed by educational researchers located in eastern Canada as a means of alerting students to the impacts of climate change. Creating an intimate link with ambient elements can, they argue, prompt student reflection and engagement: "Do we really want to lose the piping plover, the boreal clitonia, snow, the return of spring, and so on?" (Pruneau et al. 2001, 135).

For Shiva (2005, 2008), giving enhanced attention to local nature and restoring value to local sustenance cultures and economies calls for "living democracy," a de-emphasizing of (consumer-fuelled) representative and distant democracy by giving greater weight to local participatory democracy based on a keener, immediately lived, appreciation of the "interdependence between nature and culture, human and other species" (2005, 82). Opening the way to close-at-hand democracy does, however, carry the ever-attendant danger of protectionism and insularity, raising the specter of the climate change equivalent of the gated community, especially in the metaphorical North, which is why a concomitant commitment to a global justice ethic is a vital necessity. If that can be achieved, then 'denizenship' learning for conscious occupancy and participation in place, becomes an important climate change response, 'denizen' connoting primacy of immediate context while also sidestepping the built-in anthropocentrism of 'citizenship' in that a denizen can be human or other-than-human.

ADDRESSING DESPAIR, PAIN, GRIEF, AND LOSS

The piping plover and boreal clintonia may become extinct. Spring, as many know it, may become a memory. The onset of global heating is already a time of loss and will become more so, not at a metronome pace but oftentimes with bouts of breathtaking suddenness and fickleness. Breaking through the tissue of denial that marks much of the response to climate change, our chances of significant mitigation make it urgently important that we confront despair, pain, grief, and loss in our living and learning.

The counter-argument is often put that 'gloom and doom' is disabling and disempowering. Working within sustainability education circles at a university, I am occasionally invited by colleagues to special meetings to indulge in 'blue skies' thinking, by which is meant the brainstorming of positive alternative ideas for progressing sustainability. I am often chided for replying that the time would be better spent engaging in some 'dark skies' thinking that starts out from and unflinchingly confronts the dire threats we face. As I see it, the oft-stated view that positive action will only follow from a roseate frame of mind is yet another, perhaps oblique, manifestation of an untrammeled growth and progress paradigm. A more circular philosophy is needed.

For O'Murchu (2004, 190), confronting denial begins with facing up to the ecological fact that there is a dynamical unity between life and death, the "cycle of birth-death-rebirth." "We are compelled," he writes (ibid, 190–93), "to assert what seems initially to be an outrageous claim: a radically new future demands the destruction and death of the old reality. It is from the dying seeds that new life sprouts forth. Destruction becomes a precondition for reconstruction; disintegration undergirds reintegration." For Elisabeth Kuebler-Ross (1997), the terminally ill face a "grief cycle" of denial, anger, bargaining, depression, and acceptance, a cycle that is also evident in the "little deaths" or painful loss experiences of life. The multiple "little deaths" already experienced and to be experienced in the face of global heating call for a process of shedding what Kuebler-Ross calls "the winter coat of attachments that hinder new growth" (Oyle 1979, 153). Transformative learning, then, requires a conscious and thoroughgoing progress by groups and individuals through despair, into empowerment with healing and renewal. The "Great Turning," as Joanna Macy calls it, involves breaking through denial to confront the pain of the world, heroic holding actions to stop things getting worse, analysis of the structural causes of the damage wreaked by the "Industrial Growth Society," allied to the nurturing of alternative institutions and, most fundamental of all, a cognitive, spiritual, and perceptual awakening to the wholeness of everything (Macy and Young Brown 1998, 17–22). Macy's despair and empowerment work provides a powerful canon of activities and exercises for breaking out of denial to connect with the parlous state of the world (Macy 1983, 1991; Macy and Young Brown 1998). Such exercises will need to be the food and drink of education for sustainable contraction and moderation.

EDUCATION FOR RESTITUTION AND RESTORATION

But, what if all our best-intentioned efforts fall short of the mark and we pass the threshold of irreversible global heating? What if we face the actuality of Lynas' scenario (2007, 224) of human civilization contracting toward the poles in the face of the southern and northern advance of "zones of unhabitability"? By way of response, I offer two scenarios.

My first (dystopian) scenario is, strictly speaking, unrealistic in that it compresses, for the sake of underlining the point, the exacerbated intergenerational alienation that is likely to follow the onset of runaway climate change and the rapid demise and disappearance of what was familiar to earlier generations. It asks: "what do adults have to say to children when what they relate and recount out of their own experience can only be spoken of anachronistically and so is received as alien, otherworldly and irrelevant?" It draws heavily on a passage in Cormac McCarthy's harrowing post-catastrophe novel, *The Road* (2006, 163) and is an example, in extremis, of John Gray's observation (2008,

309) that "where the links between the generations are broken or the shared raiment of the common culture is in tatters, human beings will not flourish."

Scenario 1

"But we don't know these things," they protested. "It is always hot here; it is always dry here; save for scavenger crows, we see no flights of birds. Why are you showing us these pictures of plum and cherry trees in bloom, of meadow flowers, of markets full of green food? Your old books don't show the life we know. Even the words you use don't make sense to us. What did you mean just now when you said Marie "seemed full of the joys of spring?" She turned and looked at the class, and she understood once again that to the children she was an alien, a being from another age, another planet, that no longer existed. Just like their parents, much of her store of knowledge had no immediate and confirming living source. The tales she told were suspect. She could not construct for the children's learning the world that had marched northwards without viscerally reliving the loss as well. She could not enkindle in the heart of the children what had become ashes in her own soul. For them she seemed as ash. Like the schoolhouse walls, her teaching seemed to be crumbling before her very eyes.

The second scenario is ultimately more positive although set against a similar dystopian backdrop. It argues that, from within 'zones of habitability,' a long-term educational project of restitution and restoration can happen. It argues, too, both explicitly and between the lines, that the key features of education for sustainable contraction and moderation as described in this chapter, might appropriately inform that project. The climate description is guesswork in that far less seems clear about how global heating will affect the climate dynamics of the Southern Hemisphere as against the Northern Hemisphere (Liotta and Shearer 2007, 173).

Scenario 2

As was the case with the small cluster of future schools in southern Patagonia, the Lake School had a curriculum seeking to learn from the best of what had been before the big blindness and the big heating, preserving the knowledge for when, probably long in the future, the northward cooling began and people could advance outwards. The Lake village had been far enough south to escape the continuous big heat. Not that the climate was kind. There were often violent winds and rain bringing down trees and dwellings and flooding the fields, so the students had to learn disaster alertness. Sometimes a big dry heat would stretch south for a period and kill the crops, and then there would be hunger. But people could just about get by. A few far-seeing people had seen the potential and brought to the Lake School a storehouse of books of ideas and recorded experiences, written in many languages, about living gently and equitably on the earth, works that were used during Earth Restitution and Restoration sessions. The great spiritual writings of humankind had been collected, too, works that were used for Soul Restitution and Restoration classes. There were materials in abundance, too, on what had gone wrong: on blind consumerism, on disconnection from and hubristic domination of nature, on the schizophrenia of denial, on how the planet had been devastated by a deluded interpretation of the idea of progress. These were used everywhere in the curriculum. Not that much was taught in school. One of the great insights had been that second-hand learning in detached artificial places had built a culture of the head that dissected and deconstructed but that failed to speak to the heart. Teachers and students worked together most days with the Lake community, keeping alive in practical ways the sustainable ideas of the past in preparation for a sustainable future when the northward advance could begin. Everyone knew that the roots of the great heating catastrophe lay in a profoundly arrogant disregard of Gaia, a state of heart and mind that the indigenous Tehuelches of Patagonia had not been slow in intuiting upon their first fateful encounter with a different worldview and culture when Ferdinand Magellan landed on their shores.

NOTES

1. The term 'global heating' is preferred to 'global warming' to avoid the palliative effect of euphemism, but the latter is used as necessary to faithfully convey the chosen lexicon of others.
2. I am indebted to Alastair McIntosh (2008, 99) for reminding me of these two lines from a favorite poem.

REFERENCES

Adam, D. 2009. Scientists fear worst on global warming. *Guardian*, April 14, 14.

Bohm, D. and M. Edwards. 1991. *Changing consciousness: Exploring the hidden source of the social, political and environmental crises facing our world.* San Francisco: Harper.

Bonnett, M. 1999. Education for sustainable development: A coherent philosophy for environmental education? *Cambridge Journal of Education* 29 (3): 313–24.

Cannovó, P. F. 2008. In the wake of Katrina: Climate change and the coming crisis of displacement. In *Political theory and global climate change*, ed. S. Vanderheiden, 177–200. Cambridge, MA: MIT Press.

Capra, F. 1983. *The turning point: Science, society and the rising culture.* London: Flamingo.

Carson, R. 1998 (1965). *The sense of wonder.* New York: HarperCollins.

Chakrabortty, A. 2009. Forget growth: Let's focus on wellbeing and solving climate change instead. *Guardian* March 23, 25.

Charlton, N. G. 2008. *Understanding Gregory Bateson: Mind, beauty and the human condition.* Albany: State University of New York Press.

DCSF (Department for Children, Schools and Families for England and Wales). 2006. *Sustainable schools national framework*. http://www.teachernet.gov.uk/sustainableschools (accessed July 8, 2009).

Disinger, J. F. 1990. Environmental education for sustainable development? *Journal of Environmental Education* 21 (4): 3–6.

Elgin, D. 1981. *Voluntary simplicity: Toward a way of life that is outwardly simple, inwardly rich*. New York: William Morrow.

Eliot, T. S. 1963. *Collected poems 1909–1962*. London: Faber & Faber.

Eliot, T. S. 1921. The metaphysical poets. *Times Literary Supplement*, October 20.

Gadotti, M. 2008. Education for sustainability: A critical contribution to the Decade of Education for Sustainable Development. *Green Theory and Praxis: The Journal of Ecopedagogy* 4 (1): 15–64.

Gardiner, S. 2008. A perfect moral storm: Climate change, intergenerational ethics, and the problem of corruption. In *Political theory and global climate change*, ed. S. Vanderheiden, 25–42. Cambridge, MA: MIT Press.

Gore, A. 2007. *The assault on reason*. London: Bloomsbury.

Gray, J. 2008. *Gray's anatomy: Selected writings*. London: Allen Lane.

Hamilton, C. and R. Denniss. 2005. *Affluenza: When too much is never enough*. Sydney: Allen & Unsworth.

Hilmann, M., T. Fawcett, and S. C. Rajan. 2007. *The suicidal planet: How to prevent global climate catastrophe*. New York: St. Martin's.

Honoré, C. 2005. *In praise of slow: How a worldwide movement is challenging the cult of speed*. London: Orion.

Kahn, R. 2008. From education for sustainable development to ecopedagogy: Sustaining capitalism or sustaining life? *Green Theory and Praxis: The Journal of Ecopedagogy* 4: 1–14.

Kolbert, E. 2007. *Field notes from a catastrophe: A frontline report on climate change*. London: Bloomsbury.

Kuebler-Ross, E. 1997. *On death and dying*. London: Simon & Schuster.

Liotta, P. H. and A. W. Shearer. 2007. *Gaia's revenge: Climate change and humanity's loss*. Westport, CT: Praeger.

Lopez, B. 1998. *About this life: Journeys on the threshold of memory*. New York: Knopf.

Lovelock, J. 2006. *The revenge of Gaia: Why the Earth is fighting back—and how we can still save humanity*. London: Allen Lane.

———. 2009. *The vanishing face of Gaia: A final warning*. London: Allen Lane.

Lynas, M. 2007. *Six degrees: Our future on a hotter planet*. London: Fourth Estate.

Macy, J. 1983. *Despair and personal power in the nuclear age*. Philadelphia: New Society.

———. 1991. *World as lover, world as self*. Berkeley, CA: Parallax.

Macy, J. and M. Young Brown. 1998. *Coming back to life: Practices to reconnect our lives, our world*. Gabriola Island, BC: New Society.

McCarthy, C. 2006. *The road*. New York: Knopf.

McGregor, S. 2003a. *Consumerism as a source of structural violence*. http://www.kon.org/hswp/archive/consumerism.html (accessed March 15, 2009).

———. 2003b. *Consumer entitlement, narcissism, and immoral consumption*. http://www.kon.org/hswp/archive/mcgregor_1.htm (accessed March 15, 2009).

McIntosh, A. 2008. *Hell and high water: Climate change, hope and the human condition*. Edinburgh: Birlinn.

Narain, S. 2009. A million mutinies. *New Internationalist* 419: 10–11.

Okolie, A. C. 2003. Producing knowledge for sustainable development in Africa: Implications for higher education. *Higher Education* 47: 235–60.

O'Murchu, D. 2004. *Quantum theology: Spiritual implications of the new physics*. New York: Crossroad.

Oyle, I. 1979. *The new American medicine show.* Santa Cruz, CA: Unity.

Petrini, C. 2007. *Slow food nation: Blueprint for changing the way we eat.* New York: Rizzoli.

Petrini, C. and G. Padovani. 2006. *Slow food revolution: A new culture of dining and living.* New York: Rizzoli.

Plumwood, V. 1993a. *Feminism and the mastery of nature.* London: Routledge.

Plumwood, V. 1993b. Nature, self, and gender: Feminism, environmental philosophy, and the critique of rationalism. In *Environmental philosophy: From animal rights to radical ecology,* ed. Zimmerman, M., J. Baird Callicott, G. Sessions, K. J. Warren, and J. Clark, 291–314. Upper Saddle River, NJ: Prentice Hall.

Porritt, J. 2005. How capitalism can save the world. *Independent Extra,* November 4, i–vii.

Postel, S. 1992. Denial in the decisive decade. In *State of the World 1992: A Worldwatch Institute report on progress towards a sustainable society,* ed. L. R. Brown and L. Starke, 3–8. New York: Norton.

Pruneau, D., L. Liboiron, E. Vrain, H. Gravel, W. Bourque, and J. Langis. 2001. People's ideas about climate change: A source of inspiration for the creation of educational programs. *Canadian Journal of Environmental Education* 6: 121–38.

Rohr, R. 2004. *Adam's return: The five promises of male initiation.* New York: Crossroad.

Sauvé, L. 1999. Environmental education between modernity and postmodernity: Searching for an integrating educational framework. *Canadian Journal for Environmental Education* 4: 9–35.

Selby, D. 2006. The firm and shaky ground of education for sustainable development. *Journal of Geography in Higher Education* 30 (2): 351–65.

———. 2007a. As the heating happens: Education for sustainable development or education for sustainable contraction? *International Journal of Innovation and Sustainable Development* 2 (3/4): 249–67.

———. 2007b. Reaching into the holomovement: A Bohmian perspective on social learning for sustainability. In *Social learning towards a sustainable world,* ed. A. E. J. Wals, 165–80. Wageningen, The Netherlands: Wageningen Academic.

Shiva, V. 2005. *Earth democracy: Justice, sustainability and peace.* London: Zed Books.

———. 2008. *Soil not oil: Climate change, peak oil and food insecurity.* London: Zed Books.

Smith, D. C. and T. R. Carson. 1998. *Educating for a peaceful future.* Toronto: Kagan & Woo.

Tart, C. 1986. *Waking up: Overcoming the obstacles to human potential.* Boston: Shambhala.

Traina, F. and S. Darley-Hill. 1995. *Perspectives in bioregional education.* Troy, OH: North American Association for Environmental Education.

Victor, P. 2008. *Managing without growth: Slower by design not disaster.* Cheltenham: Edward Elgar.

WCED (World Commission on Environment and Development). 1987. *Our common future.* Oxford: Oxford University Press.

Worth, J. 2009. Is the economic crisis going to be the end of green? *New Internationalist* 419: 1.

WWF-UK. n.d. *Climate chaos: Learning for sustainability.* Godalming: WWF UK.

Zohar, D. 1990. *The quantum self: A revolutionary view of human nature and consciousness.* London: Bloomsbury.

3 Peace Learning
Universalism in Interesting Times

Magnus Haavelsrud

INTRODUCTION

Against the background of the pioneering work of those who have contributed towards the definition of the field of ecological security (cf. Falk 1971; Mische 1997 and 2006; Kvaløy Setreng 1973 and 2001; NCLIE 2005; Verhagen 2006), this chapter purports to take stock of the educational task of developing solidarity with nature as part of global citizenship. I use the concept of 'universalism,' which means that this would require of the world's peoples to search for a vision of commonality. It is imperative that this unity is embedded in a philosophy of conflict transformation supporting a dynamic—neither stagnant nor monolithic—universalism. I argue in this chapter that this dynamism needs an educational approach.

I view peace learning as a tool in the construction of a new universalism to be developed from below. This tool is in line with a proposal by Kumar d'Souza (1985) calling for a people's paradigm in which the oppressed peoples of the world participate in the transformation of their own societies and participate in the construction of a new world order. She points to the need to find new words, to envision new ways, to develop a worldview that would enable us to go beyond the existing orthodox models and classical paradigms and create out of "the materials of human spirit something which did not exist before." This alternative vision cannot be defined by any political dogma but will be born out of the struggles of the people in a changing historical reality: an alternative that in its dynamic possibilities can respond creatively to historical events; an alternative whose criteria would be the satisfaction of human needs and aspirations and not the hegemonic preservation of power. She argues that the history of whole cultures has been reduced to economic categories of growth and progress based on the idea that they should catch up with the standard of the industrialized nations in the North.

This new universalism, d'Souza points out, should not accept the imposition of any monolithic, "universal" structure under which it is presumed that all other cultures must be subsumed. A new universalism will have to respect the plurality of different societies, of their philosophies, their ideologies, traditions, and cultures. Movements for change need to develop conceptual paradigms born of *praxis* rooted in the specificity of social, cultural, and political processes.

The task of developing a new universalism in which all cultures participate in a more equitable way is taken a step further by Odora Hoppers (2002) arguing for an integration of indigenous knowledge systems and scientific knowledges. She regards this as a matter of cognitive justice (Visvanathan 1997 and 2002) in a world where the greater part of peoples' knowledges are subjugated and not recognized as valid. In my view such recognition of subjugated epistemologies will have a profound effect on the construction of a new universalism. The neoliberal conception of the human being as a strategic actor (see, for instance, Becker 1976) will then be confronted with conceptions recognizing human beings as capable of solidarity, community, and altruism. The human being as a strategic actor seeking profits and opportunities for himself or herself without much thought for the local community or for the world is a stranger in most cultures. The human being many of us have experienced is a person who is not reduced to a calculator of profits and opportunities but a person who enjoys a community, a world, and a universe in which continuity rather than fragmentation and discontinuity is the hallmark. This does not mean that the human being is not *also* an economic thinker and doer. I argue however that she or he is not reduced to the economic way of looking at life (Becker 1993).

PEACE LEARNING AS PRAXIS

Peace learning as *praxis* is inspired by a concept of knowledge that integrates reflection and action. This type of peace learning is transformative in the sense that both learners and contextual conditions are changed as a result. The content of peace learning is of relevance to the learners in the sense that their subjectivities are taken into account in the decisions about which content to pursue. It is problem-centered in that the learners' hopes, aspirations, and dreams in combination with a critical view of reality inspires the reflection about why reality is like it is, why it has not been changed, how future reality will be if no change occurs in present trends, which alternative futures could be considered, which one would be the most desirable, and which is more possible to realize. Changing everyday practices to a *praxis* in which reflection and action are integrated would also produce knowledge about tactics and strategies of how to change from a problematic present reality to a less problematic future. These knowledge components relate to *diachronic* relations along the time dimension focusing the historical, chronological, or longitudinal perspectives.

In this chapter, however, I shall focus the discussion on the synchronicity along the *space* dimension arguing that actions at any point on the space axis may or may not be coordinated with changes also in other parts of reality.[1] Answers to this question of synchronicity are highly relevant for guiding decisions about the most productive actions towards transformation. Considering change at any time point—present or future—needs to take this synchronicity into account. It is therefore decisive in any thought about peace—including ecological security—to clarify the assumptions made about these relationships between levels on the space dimension. Towards the end of this chapter I shall suggest that some locations on the space axis may be more important than others as 'growth points' (Howard Richards, personal communication) for transformation.

Freire (1972) introduced the concepts of 'thematic universe' and 'generative theme' as tools in understanding the relationship between macro conditions and everyday life. These concepts are useful for our discussion of the question of synchronicity along the space dimension. What is the thematic universe of our time? It seems that Freire's description in the 70s is still valid in spite of the fact that some oppressive realities have been transformed since that time. If we look at the agenda of the United Nations General Assembly over the years we still see the classical items such as eradication of poverty, disarmament, human rights violations, eradication of nuclear weapons, and so on. But a new item is also on the agenda: climate change (Mische 1997). As laid out in the Introduction to this volume, there is no doubt that if present trends are not reversed, within the foreseeable future densely populated lands will be under water or otherwise uninhabitable forcing large numbers of people to migrate to safe areas. From this thematic universe in which the 'old' problems of poverty, disarmament, and human rights violations are combined with permanent natural disasters of such magnitude, it is not difficult to imagine *generative themes* related to sheer survival in such radically changing natural conditions.

Comparing generative themes in thematic universes in which natural conditions are stable and when they are not, a qualitative difference appears. Confronting problems and conflicts that are human-made assuming that nature shall remain the same is a less challenging task than when the same problems and conflicts have to be tackled in combination with radical climate changes. In the first instance it is the contradiction between the oppressed and the oppressor that is the core of potential transformation whereas in the second instance all people face the same problem—even though people with means will be more able to survive and adapt than people with no means. Another qualitative difference between the two thematic universes is that *time* becomes a more important dimension when it comes to preparing for tackling the problem of climate change in order to survive. Nature does not postpone its disasters. Human-made disasters may at least be survived over long periods of time by those who are strong enough to deal with inhumanly miserable conditions generation after generation.

Just as contextual conditions will influence the content of peace learning as *praxis* this praxis becomes the tool for the construction of new contextual conditions and/or the transformation of the old. The dialectics between peace-learning content and contextual conditions is therefore not limited to existing conditions as long as those conditions are changeable. However, natural contextual conditions such as the climate cannot be influenced in the short term. Nothing can be done overnight with the weather. Immediate actions may have long-term effects but immediate effects are not to be seen. This is different from confronting human-made problems that sometimes may have some immediate solutions. Some effects of transcending conflict behavior may be immediate. And some remedies for solving practical matters in everyday life may occur instantly, exemplified by digging a well to deal with water shortage. Solutions to the problems incurred by climate change require long-term planning. Short-term solutions are easier to find in dealing with nonclimate problems. This poses the interesting question of how different conceptions of time relate to finding effective solutions towards confronting climate changes. The frame of this chapter does not permit a discussion of this in spite of the fact that answers to this question are highly relevant for finding answers to which growth points would be more likely to produce changes towards lower CO^2 emissions. A hypothesis may be that it may be difficult to expect from those who struggle for daily survival to prioritize such long-term strategies.

As pointed out in the Introduction to this book, it is the clear opinion of experts that the climate has been influenced by high levels of CO^2 emissions and a reduction of such emissions can reduce global warming in the long term. The content of peace learning, therefore, needs to take both short-term and long-term remedies into account. This means that in the new thematic universe in which the 'old' oppression is combined with natural contextual conditions such as climate change, it becomes imperative to locate generative themes that inspire and even demand long-term strategies for action and not only short-term tactics aimed at improvement and transformation of everyday life. The two are not mutually exclusive and can easily be combined. But when some environmentalists suggest changes in behavior in everyday life they oftentimes base their ideas on the proverb 'many small creeks will make a big river.' These changes in everyday life are important, but they are more important in the lives of those people who have big (many) houses, who have big (many) cars, than in the lives of people who have a small (or no) house and a small (or no) car. The river will increase more if those who consume most change their behavior. To develop strategies for change at the consumer end seems inadequate in that the root cause of global warming can hardly be limited to the problem of consumer behavior. The rules of production and marketing cause the kind of goods and services that contribute to high emissions—not consumer psychology. It may be argued that consumer demand causes the supply. But this argument is only valid as part of specific economic theories and has no universal validity in spite of the fact that powerful political forces argue that it has.

Conspicuous consumption is not for everyone and it seems that these strategies of limiting emissions, such as changing to energy-saving light bulbs and adding insulation to walls, really do not get at the root cause of global emissions. At the same time these tactics are not relevant as generative themes for all those people who suffer misery and oppression and whose subjugated knowledge has not contributed at all to the dominant rules of production, economic growth, market ideology, and neoliberalism. It seems that environmentalists who emphasize changes in consumer attitudes but leave the rules of production and marketing untouched have chosen this angle for change either because they believe that it is actually consumer attitudes that will spearhead the change in rules of production and marketing or because they believe that the problem itself is solely attitudinal and restricted to the public's dispositions. In an appearance on the Ophrah Winfrey Show in the fall of 2008, Al Gore focused on the behavioral dimension and less on the structural dimension in the quest for lower emissions. Such a focus may be seen as an attempt at getting at changes in the structural rules by creating a powerful enough consumer movement so that the corporate structure has to adapt to the wishes of this movement. This may be an idealistic venture as long as the rules of capitalism remain intact and without challenging centers of corporate and political power. In a recent initiative a campaign to repower America with 100% clean electricity within ten years was launched (Repower America). This is a bold and important proposal which would lower emissions considerably and serve as a model for other countries. But it does not question the dominance of contextual conditions such as neoliberalism, conspicuous consumption, or market and growth ideologies.

DISPOSITIONS AND CONTEXTUAL CONDITIONS

An essential part of a new universalism relates to our understanding of the dialectics between dispositions and contextual conditions. This understanding is an important base on which to determine effective pathways towards any kind of transformation. The call for the multitude of voices from indigenous knowledge systems made earlier would contribute to the richness in understanding how the human being's attitudes, cognitions, and volition relates to contextual conditions. A new universalism therefore would be a product of the insights from all these knowledge systems as well as their dialogical encounter with each other. This very process would contribute towards the recognition and activation of subjugated knowledge systems and therefore create a different set of contextual conditions in which both contents and forms in peace learning would be colored by this new plurality.

The basis for this call for peace learning as *praxis*, however, is exactly the belief that cognitive justice is not only the most ethical way forward but that such peace learning is rooted in the essential idea in Bourdieu's (1984) work that individual dispositions and objective reality or contextual conditions seek harmony. This force towards harmony implies that when the dispositions are stronger than the contextual conditions, the latter is transformed to fit the dispositions. Vice versa, when the contextual conditions are stronger than the dispositions, the latter would have to adapt to the former. Both are possible depending upon the relative strength of dispositions and conditions. This force towards harmony also tells us that simultaneous change in both dispositions and conditions renders another alternative in the transformation process (Haavelsrud 2007). It is therefore necessary to consider these three alternatives towards the construction of a new universalism in the interesting times ahead—confronting the inconvenient truth of climate change—utilizing the tool of peace learning.

Objective reality or contextual conditions are concepts that can be broken down in various ways. One way of classifying would be to talk of the social, political, economic, and cultural. The history of social science shows that these parts have often been researched in isolation, which has oftentimes led to fragmentary views of contextual conditions. In recent decades we have seen a scientific development in which these parts are seen more in relation to each other in interdisciplinary and transdisciplinary research institutes and journals.

One focus has been on the relationship between the cultural, the economic, and the political. This is evident in Webb's (2008) discussion of Klein's (2007) thesis that neoliberalism has gained such a dominant position in the present economic order because of a shock doctrine involving violent repression of oppositional forces. Referring to Richards and Schwanger's (2006) analysis of the rules of capitalism, Webb argues that the cause of the rise of neoliberalism is not explained by political shock treatments by neoliberal governments in power, but rather a culture disposed towards accepting and supporting the rules of capitalism. With these rules as part of the contextual conditions it is only logical that such economic rules produce and require a neoliberal economic system. We see here that by analyzing culture and economics in relation to each other a different explanation results than when analyzing the political and the economic parts without reference to the cultural. Webb is concerned that Klein's explanation focusing on political repression will alienate people from action to transform neoliberalism towards alternative economic systems developed from below at the community level. And he finds Richards and Schwanger's explanation of the neoliberal failure to produce justice and equity to be more useful in the building of an alternative economic system because of their emphasis on the cultural level. I support Webb in this analysis as I believe that no economic system can exist in the long run without cultural support. The force towards harmony between objective reality and dispositions mentioned earlier is also in support of this understanding of the relationship between cultural and economic realities: in case of disharmony the culture and the economy will adapt to each other. Such adaptation between culture and economics is portrayed in many of the predictions available in the field of climate change, as for instance *grass skiing* as a new leisure industry in compensating for the decrease experienced by the ski industry due to snow deterioration (IPCC 2007). Is it too far-fetched to imagine grass skiing as a new Olympic branch as a result of such new cultural and economic realities? What is missing in writings on forecasting such new adaptations is the idea that the human being is not only an object for adaptation but also a subject in creating culture. The concept of adaptation is used by many researchers in the volume just cited when they are forecasting climate changes and their consequences. This concept is too restricted for an educator whose basic assumption is that the human being is a subject—not merely an object. An educator would always look for the possibility of changing the contextual conditions—not only the human beings. Prophylactic learning and education are oftentimes missing from such forecasting, an omission influenced by positivistic methods in the natural sciences.

Klein's emphasis on the shock doctrine may be more applicable in political systems in which democratic participation is forbidden or significantly reduced compared with political systems based on a highly participative democracy. Such *political* differences in contextual conditions may be combined with differences in the social system. If the class structure is more or less permanent with large and maybe increasing gaps, it is to be expected that the high volume of structural violence and injustice will make participation difficult on the part of vast numbers of inhabitants. If the class structure is characterized by a lower volume of structural violence and injustice with greater possibility of social mobility, greater participation of all is more possible (Haavelsrud 2007). The shock doctrine may be more easily applied in the former than in the latter case. More than that: with high levels of participation it may be assumed that elected politicians are more representative of the people who elected them than when the level is low. This does not however imply that opposition to the rules of capitalism would be more likely in the former than in the latter. But it would imply that education as a tool in cultural change would be more possible than in a context in which the shock doctrine is more likely.

One conclusion that can be drawn from this line of thought is that the contextual conditions involving climate change as part of peace learning as *praxis* needs to be based in an analysis of the relationship between the social, the political, the economic, and the cultural. An argument against this view may be that such a task is too difficult for the greatest social scientist and that it is just too complicated for a teacher, facilitator, practitioner, or activist seeking transformative change of existing violence against nature. My answer to this argument would be this: how can anyone embark on peace learning initiatives without analyzing contextual conditions, as no peace learning can be separated from these conditions? Whether we like it or not: as long as someone purports to educate towards peace—including ecological security—then there will always be implicit or explicit assumptions made about how contextual conditions influence both contents and forms of this education and how selected contents and forms are contributing to building more peaceful contextual conditions. It is then better to spell out these assumptions than to hide them.

GROWTH POINTS

Our attention in this chapter is on the synchronicity between levels along the space dimension. The analysis of this synchronicity is necessary in order to find out which actors and actions would have the greater possibility of resulting in transformation from what is problematic and unwanted to what is desirable. Variations in the relationships between the social, political, cultural, and economic sectors in different societies are so profound that growth points have to be found in each specific contextual condition. In spite of the fact that all societies share a world in common, these variations are so big that growth points need identification as a result of an analysis of the specific contextual conditions in a specific society and in a specific *close reality* within that society. I am referring to the close reality in which people live their lives. The multitude of close realities relates to the larger systems in very different ways, and to prioritize growth points can only be done from within each close reality.

A structure can be defined as relatively permanent relations between given units (Mathiesen 1980, 14). This definition sees a structure as the result of a process of interaction among agents. This means that a structure depends on what actors, agents, or units do in relation to each other. Although Mathiesen refers to material structures I see no reason why this definition could not also apply to nonmaterial structures. The interactions could then be between actors in both civil and official society: individuals, social groups, classes, genders, ethnicities, and so on as well as municipalities, provinces, and states. Actors could be nongovernmental or international governmental organizations as well as corporations and small businesses. In order to find growth points it seems advisable to evaluate which interactions among which actors serve to maintain the structure and which are possible change agents. In theory all points are growth points as long as human beings are capable of agency and transformative action. We see it time and again that even people at the center of power serve vital roles in transforming that power (e.g., Daniel Ellsberg, de Klerk, Deep Throat/ Mark Felt, and Gorbachev[2]). This means that those close realities which are power centers can be important growth points for resistance and change whether or not the agents are at the steering wheel during the actual transformation—as exemplified by the government of de Klerk—or close enough to the steering wheel to observe the need for change and documenting that inside story to the public after leaving the arenas of power, such as Daniel Ellsberg. So when I stated earlier that transformation can occur only from within each close reality and from below it was not to exclude close realities at the macro level.

It is therefore important to consider the synergetic effects of selecting growth points at different levels on the space axis. Referring again to the *force of synchronicity* in which objective reality and dispositions seek harmony, this energy must be considered in any change process and in all specific contexts. Ideally, selecting growth points for changing dispositions in combination with growth points at organizational and institutional levels would increase this possibility for synergy. However, such synchronicity is not possible in all contexts so then the growth points may have to be selected at the dispositional level—and sometimes underground as in the case of so many political movements that were successful in overthrowing existing regimes, for instance, the African National Congress under apartheid. Only after the removal of the Botha government was the context right for negotiations between the forbidden and imprisoned resistance and the government of de Klerk. The ideal simultaneous change of macro structures and dispositions is therefore not possible in all contexts.

CONCLUSION

A focal point in this chapter has been the need for constructing a new universalism through peace learning reflecting on the relationship between the concrete and specific close reality of everyday life and other points on the space dimension. This reflection always works in tandem with *action* in such a way that one is not possible without the other. This inspiration from the process of everyday interactions generates the subjective knowledges about how that everyday life is part of a larger system in which social, economic, political, and cultural conditions shape everyday practice. These subjectivities transform into a *praxis* that may change both the interactions and the contextual conditions. In the present interesting times we are also confronted with a new set of conditions that previously were absent, viz., new human-made limits imposed by *nature*. Natural conditions are independent variables whereas the social, the economic, the political, and the cultural conditions are dependent. The latter must adapt to the former. When they do not there is an antagonistic contradiction between human-made systems and natural systems, in which the former is bound to lose if not adjusted or modified according to the laws of nature.

This has created another kind of thematic universe in which it is clear that human rights violations must be seen in combination with violations against nature. It is the clear mandate of education not to contribute to these violations either by omission or commission—in order to avoid contributing towards the legitimation of these violations against humans and their habitat. This means that a major aim of education is to respect and depart from generative themes in the community in order to analyze the relationship this micro reality has to contextual conditions in the macro world. Understanding these relations will enable the learners to decide on actionable growth points for change in both micro and macro conditions—be they social, economic, political, cultural, and *natural*. As education is a cultural tool it is therefore important to develop the analysis of the relationships between the cultural and the social, economic, and political. Education as a cultural instrument primarily focuses on the dispositions (including cognition, emotion, and volition). It has therefore been an important point in this chapter to argue for the synchronicity between the cultural and objective reality, realizing that they seek harmony. This force or energy gives a lot of power to education in that it might resist objective reality and possibly also change it. Such change, however, would be more possible in cases where there was some collaboration between such resistance from below and resistence from above. The synergetic effect of combined and simultaneous change in structural rules with the emergence of new interactions from below would be more effective in transformation than changes in dispositions only.

This means that selection of growth points for change of the present neoliberal market 'universalism' has to be based on an answer to whether it is the political shock doctrine or the rules of capitalism rooted in and maintained by culture that is the basis of its economic hegemony. It may be argued that this dominant economic paradigm is in an antagonistic contradiction to the necessity of lowering CO^2 emissions. Economic growth, market ideology, and capitalistic rules are then in contradiction to the imperatives of nature. As we know, the dominant economic paradigm has also produced its share of human misery not the least through its structural adjustment programs developed in the latter part of the 1980s giving nations an "offer they could not refuse" (Haavelsrud, forthcoming). It seems that peace learning as a tool towards developing a new universalism sensitive to indigenous knowledge systems as well as scientific knowledge would need to show as much interest in the dominant and alternative economic theories as the economists have shown in education. Human capital–inspired theories have received no less than three Nobel Prizes (four if we also count Milton Friedman) over the last few decades (Schulz in 1979, Becker in 1992, and Luckas Jr. in 1995). I am not aware that any such prize has been given to an educator. It is not much more than about twenty years ago that *Our Common Future* (World Commission on Environment and Development 1987) embraced economic growth in what was called 'sustainable development.' This political embrace of hegemonic economic theory has no support in science and needs critical analysis of the kind that Richards and Schwanger (2006) have undertaken on how this culture has embedded the rules of capitalism. And by combining indigenous knowledge systems with scientific knowledge it would be expected that a pluralistic and diverse set of economic thought would compose the new universalism coming out of peace learning—putting the World Bank back to school to learn about a sustainabililty that is built on alternative economics based on values of solidarity and equity as opposed to values of competition and growth.

The new universalism involves developing alternative futures in which conflicts are transformed by nonviolent means and institutions support interactions resulting in higher levels of human rights realization. This would imply that cultural institutions such as mass media, education, and research give priority to both facts and remedies of remaining human rights violations within frames set by natural conditions. This focus on human rights highlights both negative peace (absence of physical violence) as well as positive peace (presence of social justice). It therefore focuses our attention on both disarmament and sustainable development. Without this comprehensive or holistic understanding of desirable futures it will be difficult to find the necessary means towards its creation (Reardon 2008).

The methodology of designing a desirable future may be compared to the difficult task of understanding present realities. In the multiparadigmatic field of social science we find diverse and contradicting methodologies and theories on how to understand present society. A major controversy is related to the relationship between micro and macro realities and the degree of synchronicity between levels. Assumptions made about this relationship are of course decisive in any consideration of peace learning as the concept of learning refers mainly to the interactive level and the concept of peace refers to all levels including contextual conditions in the macro. Peace learning related to climate change therefore requires that the assumptions about the relationship between levels are stated. Without a clarification of these assumptions peace learning will be difficult—if not impossible—to design. In this chapter the synchronicity between the multitude of close and distant realities on the space axis has been the focus of attention for developing peace learning as *praxis* in a never-ending process of building a new universalism in which all cultures contribute within limits set by nature and human decency defined in terms of human rights.

SNAPSHOT 1

In 2020 UNESCO organizes a conference to discuss the concept of knowledge in formal educational systems. Recent research findings indicate that the systematic school failure of pupils from working class families, peasants, and other marginalized groups is caused by a concept of knowledge which makes it difficult, if not impossible, to integrate knowledge obtained in the family, among friends, on the Internet, and in the community with school knowledge mainly derived from scientific disciplines. School knowledge is expert knowledge and tends to be restricted to specializations in solving specific problems. It also restricts knowledge about the future to *forecasting* and shies away from visionary knowledge about alternative and potential futures other than the one forecast within specific areas.

The Norwegian philosopher and cofounder of the field of ecophilosophy Sigmund Kvaløy Setereng is invited to address the conference. A main thesis in his keynote address draws from his book on the *Eco-crisis, Nature, and the Human Being* (published in Norwegian and entitled *Økokrise, Natur og Mennesket*, 1973). His thesis is adopted in the *Final Document* and reads:

> Nobody can avoid becoming a philosopher because we always behave in accordance with a world view in which basic questions about existence and meaning either are answered or assumed. The search for insight into many fields of study and experience from many life situations need to be brought together in a wholeness which is based on reason, i.e. developing a philosophy. As this philosophy also is of political interest it means that it has to be turned into political action.

With this new orientation to knowledge in formal education, the equity gap between pupils from working-class and middle-class families was reduced 60% on an average by 2030 in member countries of the Organisation for Economic Co-operation and Development (OECD). A new UNESCO conference on evaluation of the effects of this new concept of knowledge on obtaining more equity concluded that the main problem with the traditional concept of knowledge had been the dislocation of knowledge from the knower—pointed out by many in the last century including the late, famous professor Basil Bernstein at the Institute of Education at the University of London.

With the new concept of knowledge the inner and the outer is integrated implying that the world views of the pupils from the working classes are respected equally with world views from more dominating classes, the evaluation concluded. Cognitive injustice arising from systematically subjugating knowledges of peripheral groups in formal education is transformed into a formal education integrating formerly subjugated knowledges with scientific knowledge. This integration of knowledges had for decades been advocated by many and from diverse cosmologies as evident in the writings of the Japanese educator Yoshiko Nomura inspired by Buddhist thought emphasizing the concept "lifelong integrated education" (NCLIE, 2005), the Norwegian peace researcher Johan Galtung (1996) integrating theories of development, conflict, and cosmologies, and the Brazilian educator Paulo Freire (1972) whose conscientization approach enabled the voiceless to speak and be heard. Other voices from what had been previously called 'developing countries' and the 'third world' also included Catherine Odora Hoppers (2005) who from her development education chair at the University of South Africa embedded the imperative of nature in her concept of Indigenous Knowledge Systems (IKS):

> . . . IKS practice does not seek to conquer or debilitate nature as a first impulse. This can be contrasted, for instance, with the counselling of Francis Bacon and the mechanistic conception of reality, which urged a 'passionately vicious' approach to nature. 'Nature' had to be 'hounded

in her wanderings, bound in service, and made a slave. She [note the pronoun] was to be put in constraint. [The work of the scientist was to] . . . torture nature's secrets from her' (Capra 1988, 226).

IKS stresses instead the essential interrelatedness and interdependence of all phenomena—biological, physical, psychological, social, and cultural. Indigenous cosmology centers on the coevolution of the spiritual, natural, and human worlds. Thus many indigenous peoples in Africa still practice the ritual of burying their umbilical cords and immediately planting trees on the spot in order to establish a relationship with plant life. Family histories make reference to some animal totem to be conserved.

IKS holds that there are sacred places that have to be avoided and must be conserved. There are places where people are not permitted to fell trees, hunt wildlife, or collect wild fruit for commercial purposes. Natural phenomena like rivers and mountains play a significant role in the psyche and constitution of communities. Experiences from indigenous communities in other parts of the world emphasize the fact that knowledge is relationship, and relationship brings with it responsibilities and obligations and extends into ecological practice (Peat, u.d., 1987).

SNAPSHOT 2

Ecophilosophy becomes a required course in teacher education in Norway in 2015. The course reflects the insights that this discipline is not value neutral and descriptive only. It is explicitly normative aiming at developing and utilizing the students' commitment towards and engagement in thinking and acting towards ecological security as part of the more comprehensive field of peace studies. In this way ecological security is conceived as closely integrated with issues of development, disarmament, and human rights. Evaluation criteria in the course include those described by Sigmund Kvaløy Setreng in his book, *Diversity and Time* published in Norwegian and entitled *Mangfold og Tid* in the year 2001 page 3271:

1. documentation of understanding global and local natural conditions seen as wholes in which each part relates to all other parts;
2. documentation of understanding that each human being and the human species are part of nature demonstrating cases of both conflict and harmony;
3. demonstrating a wide range of human abilities and possibilities such as critical thinking, understanding patterns, wholes, and processes of change, sensibility, empathy, emotion, and compassion, intuition, and not the least: ability for practical and political activity by participating in political and/or social conflict in the struggle for ecological security.

The students are required to focus on problems and challenges in their own community. One group of students in Trondheim, Norway, observes how plastic bags pollute not only beaches but oceans as well. As birds on the beaches and on the surface of the ocean get tangled up with this plastic they die and sink becoming part of the fish food chain. This can over time lead to bits of plastic in fish stomachs which again are eaten by bigger fish, seals, and whales. As plastic contains chemicals that are cancerous the prognosis is that not only animals in the oceans may be exposed to this danger but human beings will in the end find it on the dinner table. In their search for causes of this problem the students are confronted with not only human behavior on the individual level, but also political apathy in the Norwegian government and international governmental organizations such as the UN. Their goal or vision is to reduce the use of plastic in food wrappings, for shopping purposes, and also secure safe storage of plastic that is thrown away. This is a structural problem requiring global politial action and they contact the representative from Norway in the United Nations Parliamentary Assembly (UNPA) in order for her to take up this issue in this global lawmaking institution. She finds the obstacles from industry and commerce so strong that she refuses to do so. Upon this proof of more political apathy the students become furious and contact the major newspapers in the country and with the help of the Young Peacebuilders Organization in the Hague they are able to make contact with similar youth organizations in several countries about this issue. The local challenge observed on a beach in Trondheim with the help of the teacher-education students led to a global movement to forbid the use of plastic in food storage and transport including shopping bags in all member countries of the UN. This student group meets the three criteria of evaluation mentioned earlier and received a good grade in the course on eco-philosophy: (1) they had demonstrated the local-global interrelationship, (2) they had demonstrated harmony and conflict with nature, and (3) they had demonstrated a wide range of abilities and integrated reflection and action in their course work. PS: It took five years for the seed sown by the teacher education student group to flourish in a UN law. By the way the representative from Norway was not reelected to the UNPA.

NOTES

1. It is however important to integrate both diachronic and synchronic relations as significant dimensions when designing the contents of peace education (Haavelsrud 1996, chap. 2).
2. Daniel Ellsberg was a military analyst who released the Pentagon Papers dealing with the Vietnam War strategies to the press. Frederik Willem de Klerk as President of South Africa entered into negotiations with the African National Congress on ending *apartheid*. Deep Throat was a pseudonym for William Mark Felt, Sr.—he released information to the press on the Watergate scandal leading to Nixon leaving his office as President of the United

States. Mikhail Gorbachev, as the leader of the former Soviet Union, helped end the cold war and the political monopoly of the Communist Party.

REFERENCES

Becker, G. 1976. *The economic approach to human behavior.* Chicago: The University of Chicago Press.
———. 1993. *Human capital: A theoretical and empirical analysis, with special reference to education.* 3rd ed. Chicago: The University of Chicago Press.
Bourdieu, P. 1984. *Distinction: A social critique of the judgement of taste.* London: Routledge & Kegan Paul.
Capra, F. 1988. *Uncommon wisdom: Conversations with remarkable people.* London: Flamingo.
Falk, R. 1971. *This endangered planet.* New York: Random House.
Freire, P. 1972. *Pedagogy of the oppressed.* Harmondsworth, Middlesex, UK: Penguin Books.
Galtung, J. 1996. *Peace by peaceful means: Peace and conflict, development and civilization.* London: Sage.
Haavelsrud, M. 1996. *Education in developments.* Tromsø: Arena.
———. 2007. Synchronizing cultural and structural changes towards global governance: International security, peace, development, and environment. In *Encyclopedia of life support systems (EOLSS),* ed. U. O. Spring. Oxford: Under the Auspices of the UNESCO, Eolss Publishers.
IPCC. 2007. *Climate Change 2007: Impacts, adaptation and vulnerability. Contribution of Working Group II to the Fourth Assessment Report of the Intergovernmental Panel on Climate Change,* ed. M. L. Parry, O. F. Canziani, J. P. Palutikof, P. J. van der Linden, and C. E. Hanson, 541–80. Cambridge: Cambridge University Press.
Klein, N. 2007. *The shock doctrine: The rise of disaster capitalism.* New York,: Metropolitan Books/Henry Holt.
Kumar d'Souza, C. 1985. *The reality, the rhetoric: The need for a new political culture.* (Unpublished contribution to the manuscript for the UNESCO Teacher's Handbook on Disarmament Education). Trondheim/Paris: Norwegian University of Science and Technology Library Archive No. 303/UNESCO Archives.
Kvaløy Setreng, S. 1973. *Økokrise, natur og menneske: En innføring i økofilosofi og økopolitikk (Ecological crisis, nature and the human being: An introduction to ecophilosophy and ecopolitcs).* Trondheim: Videnskabs Selskabs Bibliotek.
———. 2001. *Mangfold og tid: Pyramide-mennesket ved skillevegen: System, frihet eller kaos? (Diversity and time: The pyramid person at the road crossing: System, freedom or chaos?)*Trondheim: Musikkvitenskapelig institutt, Norwegian University of Science and Technology.
Mathiesen, T. 1980. *Law, society and political action.* London: Academic Press.
Mische, P. M. 1997. Assuring ecological security for the 21st century: How can the UN's role be strengthened? *Breakthrough News,* September–November, 1–8.
———. 2006. Educating for peace and planetary community at the level of our deep humanity. *Breakthrough News,* October–December, 5–9.
NCLIE (Nomura Center for Lifelong Integrated Education), ed. 2005. *Wisdom for the survival of humankind.* Tokyo: Ichiyosha Publishers.
Odora Hoppers, C. 2002. Indigenous knowledge and the integration of knowledge systems. In *Indigenous knowledge and the integration of knowledge systems: Towards a philosophy of articulation,* ed. C. Odora Hoppers, 2–22. Claremont, Cape Town, South Africa: New Africa Books.

————. 2005. Culture, indigenous knowledge and development: The role of the university. In *Occasional Paper No 6*. Johannesburg: Centre for Education Policy Development.

Peat, F. D. 1987. *Synchronicity: The bridge between matter and mind*. New York: Bantam.

————. n.d. Blackfoot physics and European minds. http://www.fdavidpeat.com/bibliography/essays/black.htm (accessed March 2003).

Reardon, B. 2008. A pedagogy of alternatives: A peace education comment on Mark Webb's letter to Naomi Klein. *Factis Pax* 2 (1): 123–36.

Repower America. http://www.repoweramerica.org. (accessed March 2009).

Richards, H. and J. Schwanger. 2006. *The dilemmas of social democracies*. Lanham, MD: Lexington Books.

Verhagen, F. C. 2006. *Selective categorized resources list on sustainable communities, sustainable development and sustainability*. (Unpublished for participants in the Envisioning Sustaining Futures Consultation), April 21 at Community Church of New York.

Visvanathan, C. S. 1997. *A carnival for science: Essays on science, technology and development*. Oxford/Calcutta: Oxford University Press.

————. 2002. Between pilgrimage and citizenship. In *Indigenous knowledge and the integration of knowledge systems: Towards a philosophy of articulation*, edited by C. Odora Hoppers, 39–52. Claremont, Cape Town, South Africa: New Africa Books.

Webb, M. 2008. Letter to Naomi Klein. *Factis Pax* 2 (1): 137–59.

World Commission on Environment and Development. 1987. *Our common future*. Oxford: Oxford University Press.

4 Climate Injustice
How Should Education Respond?

Heila Lotz-Sisitka

INTRODUCTION

> Global warming needs a response that isn't only at the level of manag-
> ing an environmental problem to ensure the planet is just about live-
> able on in the years to come—it needs one that addresses the essential
> un-freedom, suffering and misery within the present global system.
>
> (Hoffmann 2007, 7)

The *Fourth Assessment Report* of the Intergovernmental Panel on Climate
Change (IPCC 2007) identifies Africa as one of the continents of the world
most vulnerable to climate change. Africa's vulnerability to climate change
is aggravated by the interaction of multiple stresses such as poverty, poor
governance, and weak institutions, limited access to capital (including
technology), ecosystem degradation, conflict and disasters (UNEP 2006),
and a generally poor quality of education (UNESCO 2004). The climate
injustices and exacerbating circumstances experienced by poor and weak
states today lie in the long-term historical emergence of a modern (and
increasingly global) world order framed by a hegemonic Westphalian state
system. This state system privileges exclusive, undivided sovereignty over
a bounded territory (Fraser 2008), and is known more popularly as the
'nation state' system.

In the contemporary world order, this state system is characterized
by marked differentials in wealth, power, and control over the flow of
resources, and is held in place by a system of hegemonic states, and poorer,
weak states (differentiated by terms such as 'developed' and 'developing'
or 'first world' and 'third world'). Within this framing, Africa consists of
fifty-four poorer and weaker states, and all of its countries are viewed as
'developing' or 'third world.' The interstate system that has increased its
influence in the latter part of the twentieth century is today dominated by
hegemonic states. Most of the hegemonic states have built their power and
economies on colonial intrusions, extractive policies, and a philosophy

of man's (sic) dominion over nature which spawned the industrial and technological revolutions. These developments are now being severely critiqued for reckless overburning of fossil fuels to power a particular notion of progress associated with the 'free' market system. Today, this system is marred by significant failures, characterized most notably by a state of a world at risk (Beck 1992, 1999, 2009). The net result of this development trajectory is that those countries most at risk from the consequences of climate change are ironically and paradoxically also the ones that have contributed least to climate change, making climate change a social justice issue that is global in nature.

The severity of climate injustice is visible in the predictions that African states will lose food production capacity, while they will be faced with increased health risk and water stress (among other severe impacts such as loss of biodiversity, heat stress, and so on). Predictions are that agricultural production and food security are likely to be severely compromised by climate change, with projected yield reductions of up to 50 percent and crop net revenue reductions of up to 90 percent. Smallholder farmers are likely to be worst affected, which gains significance in the light of the fact that smallholder farmers provide most of the food produced and used in Africa (UNEP 2006). Water stress will be aggravated in some countries, while others currently not at risk of water stress will become at risk, possibly affecting up to 250 million people by 2020 and 600 million by 2050 (UNEP 2006; Africa Geographic 2007). Southern Africa, already a water-scarce and food-insecure region will be affected. Additionally, the spread of malaria will increase (Africa Geographic 2007), and precious biodiversity will be lost with economic and livelihood consequences. Consequently the southern African region has been classified by the United Nations Environment Programme as being highly vulnerable with low coping-capacity (UNEP 2006).

These social justice issues are receiving attention in climate change science and activism (e.g., Schipper and Burton 2009; Worth 2009). One of the most hotly debated justice-related issues at the moment is the biofuel colonization of rural African lands (Havnevik 2007), which has resulted in the increase of food prices (maize prices have risen by two-thirds in the last few years). Havnevik (2007) explains that the way in which poverty, consumption, and climate change are being addressed (i.e., as large-scale economically oriented development interventions) tends to blur historical, structural, and power features underlying global inequalities. For example, Havnevik (ibid. 4) states in a critique of World Bank policies and practices in the agricultural sector: "Farmers' economic and social choices are highlighted before probing the central issue facing Africa's rural dwellers, namely the increasing displacement of their agrarian labour." Similar problems are evident in climate change responses, hence the epigraph from Hoffman (2007), which argues for a wider, social justice response to climate change in Africa and elsewhere.

EDUCATIONAL RESPONSES IN COUNTRIES AT RISK

While global and local discourses of mitigation and adaptation are arising in response to the severity of climate change impacts, there is, as yet, little material available on how countries affected by climate injustice might develop and deepen educational responses to climate change. UNESCO (2009) recently produced a policy dialog paper on climate change education (which I authored) proposing that climate change education should seek deeper understandings of mitigation and adaptation responses so that these might be socially transformative and not simply technical, and that climate change education should also strengthen practice-centered social learning approaches so that climate change education may harness creativity and be empowering and thus move beyond awareness of the scientific facts about climate change (Lotz-Sisitka 2008).

In this chapter I take this work a little further, by asking how educational responses can be oriented towards deepening critical engagement with the questions of climate injustice. My interest is to probe how such an emancipatory orientation might interface with practice-centered social learning approaches that are increasingly becoming popular among environmental educators in southern Africa who are favoring capability-centered, agency-based orientations to learning and adaptation (e.g., O'Donoghue and Fox 2009; Taylor 2009; Lotz-Sisitka 2008; SADC REEP 2009) (I include myself among these). In the past few years there has been much attention given to social and situated learning in environmental education (Wals 2007; O'Donoghue 2007), and this has ushered in a thesis that cultural-historical approaches to learning that take account of and allow for reflexive engagement with contradictions and tensions has more to offer the field of environmental education than awareness-raising and behavior-change approaches. This is because such theories of learning are also agency-centered, and have the potential to enhance capabilities and social change processes (O'Donoghue 2007), and because they value engagement with conflict and contradictions, and are also pluralist and open-ended (Wals 2007). Support for such approaches has meaning in a context where educational approaches can have material benefits for those living in poverty, especially when they can assist with the emergence of new and innovative adaptation practices (which are critical for responding to the scenario sketched out earlier). In a recent southern African consultation involving more than 600 educators from fourteen countries, there was consensus that environmental education and education for sustainable development in the region should be oriented towards involving people in sustainable development actions (Lotz-Sisitka et al. 2006).

With the potential material value of strengthening adaptation practices in a high-risk environment, such theories of learning easily become contextualized, and ironically may run the risk of becoming bounded by conservative forms of contextualism, while seeking to be emancipatory. It is this issue that I examine through a critical review of the Southern African Development Community (SADC) Regional Environmental Education Programme's (REEP) emerging educational response to climate change issues in southern Africa. I do this by drawing on the insights of social theorists and activists working at the cutting edge of social justice theory and climate justice issues. I am currently working with the SADC REEP to develop orientation and monitoring processes to guide the reflexive management of the program, hence the focus and interest of this chapter. It is in a new phase of development, and climate change has been identified as a subregional priority area for cooperation at a subregional level, influencing the agenda of the SADC REEP.

SCENARIO 1: THE SADC REEP'S CURRENT EDUCATIONAL RESPONSE TO CLIMATE CHANGE

The SADC REEP is one of the *premier* environmental education programs in southern Africa. It serves fourteen member states of the Southern African Development Community, which is a regional economic community oriented towards enhancing the development opportunities and wellbeing of the 251 million people that live in the region, 75 percent of whom are located in rural areas, where they live closely off the land. The program was established in 1993 by SADC states under the SADC Treaty. The program activities are guided by the objective of "enabling environmental education practitioners in the SADC region to strengthen environmental education processes for equitable and sustainable environmental management choices" (SADC REEP 2002). Since 1997 the program has provided various training programs, networking opportunities, and policy and materials development programs for environmental educators across the subregion. It holds a database of environmental education materials. It supports regional cooperation and educational research into environmental education responses to critical environmental issues such as soil degradation, water stress, loss of biodiversity, poverty, loss of ecosystem services, and climate change. Climate change has been on the agenda of SADC countries for a number of years, but it is only recently that it has risen to a more prominent priority issue within the SADC REEP. For a number of years, the program has explicitly supported a contextualized, social learning orientation to agency and capability development in their educational approaches as shown by this statement:

> The SADC REEP and its partners have developed an approach to environmental education that supports individuals, communities and institutions in southern Africa to strengthen alternatives and capabilities that respond to critical environmental and sustainability issues and risks. Ultimately the program seeks to strengthen socio-ecological

resilience and sustainable development in the southern African region. To do this, the program uses an inclusive, partnership and networked approach, and emphasises learning together for a sustainable future. It foregrounds transformative learning, as well as mainstreaming of environment and sustainability education into education systems in southern Africa, within a regionalised co-operative framework. (SADC REEP 2009, 1)

Within this framework the SADC REEP's orientation to climate change education is still evolving, and there is a question as to whether climate change education in the program should be any different to other foci for environmental education, and whether climate change is just a 'distractor,' taking attention from more important issues such as HIV/AIDS or rural livelihood development. To understand better what *is* already being done (as Scenario 1), I briefly document evidence of educational research, knowledge resources, and pedagogical resources available to the program for climate change education. Table 4.1 is based on a review of the SADC REEP database, a review of some early research projects related to climate change education generated by or in association with the program, and a collection of pedagogical resources easily accessible at the SADC REEP. It focuses on only those resources and educational artefacts that are *directly* oriented to climate change, and not the myriad of resources, research outputs, and materials that exist for responding to the wide variety of associated issues such as soil degradation, loss of biodiversity, water stress, and so on. It also excludes e-learning resources, which may explain the dated nature of the information resources.

The pedagogical responses to climate change outlined earlier appear to be the strongest characterizing feature of the SADC REEP's response to climate change, since a range of different methodological orientations and materials exist for this work, and new methodologies are being researched, and tried out. While a variety of approaches appears to exist, the research direction and most recent pedagogical developments are reflecting social learning and situated learning approaches that are increasingly being focused on local and small-scale adaptation practices and alternatives (Lotz-Sisitka 2008; O'Donoghue and Fox 2009; Taylor 2009). As shown by the analysis of SADC REEP resources, some pedagogical activities are also oriented towards technical knowledge transfer and individual responsibility and behavior change. Of interest in the analysis is the insight that the politically motivational materials share good examples of advocacy and awareness-raising work, but fail to engage social learning orientations or sustainability practices. Another point of interest is that while there are available knowledge resources on the issues of climate change at the SADC REEP, these appear to be minimal, and somewhat dated. The few informational materials that were found in the resource bank were mainly technical in orientation, focusing on technical environmental management–related

Table 4.1   Knowledge Resources, Pedagogical Resources, and Research Directly
Addressing the Topic of Climate Change at the SADC REEP

Knowledge resources	Of the holding of some 2400 information and pedagogical resources collected from across the SADC sub-region and catalogued at the SADC REEP for use by visiting environmental educators, only five were found to explicitly address the topic of climate change in southern Africa. Two of these resources are sub-regional resources developed by the SADC to share information on climate change across the region. The other three were written in Zimbabwe (one of the fourteen countries in SADC). The two regional resources are dated technical reports on projected impacts and implications of climate change in the SADC region, being variously produced in 1992 and 1995. These discuss impacts of climate change, regional scenarios, and sectoral policy implications for the SADC region. The two Zimbabwean studies focus on specific climate- related issues and scenarios, including vulnerability and adaptation of maize production and mitigation options related to energy technologies. Both were produced in 2004 through a funding partnership with the Netherlands government. The other (fifth) resource collected from the resources bank is a 'Fact Sheet' based on the 1990 IPCC Working Group 1 Report, thus also being dated. There are other resources such as the *South African Environment Outlook Report* (RSA 2007) and the UNEP *Africa Environment Outlook* (UNEP 2006) that carry information on climate change, but within a wider integrated framework discussing environmental issues and risks more broadly, and broader policy responses and scenarios. A sub-regional Environmental Outlook Report does not exist, except in an older version produced in 1994.
Pedagogical resources	A number of pedagogical resources are emerging to respond to climate change. A sampling of these from the SADC REEP offices indicates the availability of a number of diversely constituted resources that focus on: • *Individual and community-based behavior change*, such as a faith-based pamphlet focusing on "Why should I as a person of faith be concerned?" These and other similar materials propose a lighter footprint as the solution at home and in the community, emphasizing individual commitment and action and behavior change. • *Information and concept clarification* such as booklets and worksheets that focus on "what is climate change?" Such materials generally outline what global warming is, how climate changes are measured and predicted, and what the effects of climate change are. They tend to end with a "what can you do?" section emphasizing individual and community action, lighter footprints, and behavior changes. Such materials focus on sharing concepts, vocabulary, and facts about global warming, and are 'content rich' in a technical sense. • *Ideological framings of solutions*, such as a poster produced by the national energy company (distributed to all Eco-Schools) presenting a proposal that nuclear energy is the solution to climate change (with little space for debate or deliberation on this proposed solution).

(continued)

Table 4.1   Continued

Pedagogical resources	• *Activity-centered mediations of content and concepts* involving choreographed participatory activities, an example being a teaching pack produced by the national energy company, this time promoting learning about different renewable energy options, and also sharing information on how coal mining damages the environment, and information on climate change, greenhouse gases, and global warming (making the distinctions clear). The materials propose a number of participatory activities to involve children in the learning process, drawing on constructivist learning theory. • *Political message and motivational* materials such as those found in a resource produced by the United Nations Environment Programme focusing on the topic "towards a low carbon economy" for World Environment Day. This resource presents a number of politicians from different countries stating that everyone needs to get involved; it outlines twelve things to do, and then shares examples of exciting environmental education projects from around the world, most of which are raising awareness, and contributing little to the actual practical changes necessary for a low-carbon economy. • *Practice-centered activities and materials* oriented towards seeking out solutions and alternative practices which demonstrate or provide guidance on how to do things differently. A set of 'Handprint' materials (with opposite iconography to footprint materials) encourage positive actions for change at the local and community level within a situated learning methodology that is socio-culturally located and emergent (e.g., O'Donoghue and Fox 2009). These materials encourage the use of knowledge resources to broaden understanding of the local practices being implemented. A number of these practices are 'in operation' at the SADC REEP, and serve as demonstration sites for sustainability (e.g., use of a solar cooker to make tea; use of low carbon printing technology), although their use appears to be inconsistent (Taylor 2009). Adaptation practices being focused on include: local greening projects, local carbon sequestration activities, energy saving, introduction of low-cost alternative printing technologies, and sustainable agricultural practices. Most prominent were the practice-centered social learning activities and materials, the individual and community-based behavior-change materials, and the activity-centered mediations of content and concept materials. Practice-centered social learning activities and materials are currently receiving the most attention in terms of 'new directions' for education and learning in the SADC REEP learning environment.
Research and training resources	The research and training resources (a draft research paper, a thesis, and course materials) reveal an interest in supporting and engaging people in understanding risks (i.e., a risk epistemology), and in sustainability practices. The research interests seem to be focused on how people learn in contexts of risk where new sustainability practices are emerging, thus the research is oriented towards social learning and risk responsiveness, as well as towards capabilities and agency, although this focus is currently still developing.

aspects of climate change and policy options. Missing in the collection of educational artefacts available for climate change education at the SADC REEP were the following:

- up-to-date knowledge and information on climate change at country and subregional (SADC) levels,
- up-to-date analysis of climate change policy developments across the southern African region, and
- up-to-date analyses and methodological guidance on dealing with social justice perspectives on climate change (beyond an interest in adaptation practices).

Up-to-date knowledge resources and tools for wider political and critical engagement with climate change issues appear to be missing from the SADC REEP's climate change education resource base, while tools for situated and contextualized learning processes are emerging rapidly with the wider intent to strengthen social learning and capability-centered approaches to climate change education in a southern African context. This scenario therefore influences the available learning opportunities on climate change at the SADC REEP, which is discussed in more depth following, with openings provided for a new scenario to emerge.

Within the emerging frame for climate change education outlined in the scenario earlier, it would be typical for the SADC REEP to, for example, provide support to agricultural extension officers to develop educational programs with farmers that recover indigenous knowledge of sustainable agricultural practices and extend their strategies for companion planting (among many other such processes), or to work with school teachers and children to implement various school gardening practices, energy saving projects, water harvesting activities, and so on, in school improvement projects that also address climate change issues through active learning processes of identification of local issues, inquiry, action taking, and reflection.

IS SITUATED SOCIAL LEARNING AN ADEQUATE RESPONSE TO CLIMATE INJUSTICE?

Anne Edwards (2005) and Harold Glasser (2007) have both noted that there is nothing transformative about social learning processes *per se*, since social learning processes can be inherently conservative if oriented only towards localization and contextualization, and can ironically serve to maintain the status quo while appearing to be transformative. A focus on social learning processes *per se* does not necessarily address wider structural injustices that create the problems that agents are responding to through enhanced learning at a localized level. The epigraph from Hoffman (2007) at the start of the chapter draws attention to the breadth and depth of the climate change *problematique*. It outlines the global nature of the response needed to address climate change, and frames climate change as a global social justice issue, providing a holistic, social change framework for such climate change educational work. I consider the nature of climate injustice in more detail later, before returning to the educational implications, as they pertain to the SADC REEP, where I articulate 'Scenario 2' or an extended possibility for social learning and climate change education.

Climate Injustice

In a recent article entitled "Power Politics," Jess Worth (2009, 3) writes that climate injustice involves "a triple injustice":

- First, climate change affects the poorest first and worst. Nearly all of the climate change casualties (deaths), so far, have been among those at most risk—poor people living in what she calls the 'Majority World.' This death toll is predicted to rise to millions in just a few decades.
- Second, those most affected did not cause climate change, and are powerless to stop it. This marks out the power-gradient between industrialized nations with high fossil fuel burning economies, and those that are less industrialized (lower fossil fuel burning economies).
- Third, the polluters are not paying. Despite interstate agency interventions, commissions, agreements such as the Kyoto Protocol, the gaze of the international media, and wide-scale public awareness, carbon dioxide emissions (which account for 80% of greenhouse gasses) continue to rise in developed countries. The clean development mechanism has by and large failed to provide the support necessary for clean development in developing countries, and the carbon trade market has been plagued by mismanagement and ineffective implementation, and has ". . . perversely enriched the most polluting European energy companies to the tune of billions of dollars, and exported dirty development schemes to the global South." (2009, 6)

She reports further:

> The G8 have so far pledged a shockingly inadequate $6 billion—to be disbursed through World Bank loans, . . . forcing affected countries to pay twice for their own suffering, with the added slap in the face of stringent World Bank conditions. Compare this with the hundreds of billions chucked at bailing out the banks, with minimal conditions, and the injustice gets gut-wrenching. (2009, 6)

Worth's (2009) paper points to a failure of the Westphalian state system and its associated interstate mechanisms (such as the UN and the World Bank) to adequately address climate injustices, and points out the problem of *incommensurability*, which is also raised by Sunita Narain (2009, 11), an environmental scientists and activist who notes:

> There is a fundamental difference between the rich and poor's response to climate change. The environmental movements of the rich world emerged after periods of wealth creation and during their period of waste generation. So they argued for containment of the waste, but did not have the ability to argue for the reinvention of the paradigm of waste generation itself . . . . It is ironic that, despite the science telling us that drastic reductions are needed, no country is talking about limiting its people's consumption. Yet efficiency is meaningless without sufficiency.

The SADC REEP climate change education support materials and research (as identified and analysed in Table 4.1 and in the discussion earlier), shows little evidence of in-depth critical engagement with this complex multilayered assessment of climate justice issues at the macro level, beyond the focus on a risk epistemology in the research. To simply promote these issues as outlined earlier with activist logic may be too simplistic a response to represent the climate justice perspective for inclusion in educational programs. A deeper understanding of the changing context of justice itself is necessary.

Changing Concepts of Justice

Nancy Fraser (2008), a social justice theorist, argues that global problems such as climate change are fundamentally changing concepts of justice, which hitherto have been framed primarily by the Westphalian state system, and what Benhabib (2006, 2) refers to as the "positive law" of the nation state. She argues that claims of redistributive justice bounded within state-level justice systems are no longer adequate for dealing with the complex nature of social justice issues such as climate change. Consequently, the Westphalian justice system is now in dispute as social movements across the globe coalesce in their critiques of transborder injustices. Fraser (2008, 2) argues that such justice claims are also increasingly "mapped in competing geographical scales . . . for example, when claims on behalf of 'the global poor' are pitted against the claims of citizens of bounded polities." Her argument is that "the challenges posed in the present epoch are truly radical" (2008, 2), thus radically new conceptions of justice are required, since without this what will be left is a "spectre of *incommensurability*" (2008, 3, emphasis original). She argues further that such incommensurability is unable to resolve issues of redistribution, recognition, and representation. Applied to the climate change scenario sketched out earlier, such incommensurability exists at the level of redistribution, since resource flows are unequally shared and dominated among nation states, with unequal effects and risk production. At the level of recognition, the solution to the problem is being deflected to nation states when it is a global justice problem, and is being seen in terms of *efficiency* rather than *sufficiency* and *equity*. Incommensurability also exists at the level of representation, since hegemonies and power differentials exist among states responsible for, and at the receiving end of climate change injustices and risks. Fraser proposes a way forward out of this impasse with the idea of *reflexive justice*. She (2008, 73) states that:

> The idea of reflexive justice is well suited to the present context of abnormal discourse. In this context, disputes about the "what", the "who", and the "how" are unlikely to be settled soon. Thus, it makes sense to regard these three nodes of abnormality as persistent features of justice discourse for the foreseeable future. On the other hand, given the magnitude of first-order injustice in today's world, the worst conceivable response would be to treat ongoing meta-disputes as a license for paralysis. Thus, it is imperative not to allow discursive abnormalities to defer or dissipate efforts to remedy justice.

She proposes that reflexive justice should *at the same time,* entertain and strengthen urgent claims on behalf of the disadvantaged, ". . . while also parsing the meta-disagreements that are interlaced with them. Because these two levels are inextricably entangled in abnormal times, reflexive justice theorizing cannot ignore either one of them" (ibid. 73). This has implications for the way in which social learning is conceptualized within a climate justice frame, since it requires not only situated learning engagements focusing on the emergence of creative and materially significant adaptation practices, but also sophisticated *deliberation* and *reflexive* engagement with climate change justice questions that span the local/global and present/future time-space configurations. For example, it requires climate change activists, practitioners, and negotiators to continue to argue for, and to seek out the means of foregrounding sufficiency and equitable sharing of the global commons, for putting in place massive investments in renewables, for the systematic closing down of coal-fired power plants, for the revitalization of cheap public transport and a rapid curbing of aviation, for keeping fossil fuels in the ground (i.e., halting climate change at the supply side), for halting current forms of exploitative carbon trading, and for making a shift away from the pursuit of mass production and consumption, in short charting the path to a zero-carbon economy *while* insisting on adequate support for adaptation (Worth 2009; Narain 2009; Bond 2009; Chivers 2009). Climate activists in the global South argue that simply settling for adaptation investments in climate negotiations is not enough, there is a need to insist on charting a global path towards a zero-carbon economy *as well* (Chivers 2009). Engaging critically across these foci, with the discourses of *sufficiency* and *equity* as guide, has the potential to develop a deeper, more reflexive understanding of the nature of climate change impacts and solutions.

Fraser (2008) states that the 'subaltern' (in this case those most at risk from climate change) *need to be able to speak in authoritative terms,* while the complex aspects of the 'who,' the 'what,' and the 'how' are being worked out through changes to the current Westphalian justice system, which she and other social justice theorists such as Seyla Benhabib realize will not be easy. Narain (2009) too urges that there is a need for reflexive justice among all involved to put forward a framework for an effective climate agreement *for the entire world.* She states that such a framework must be based on two imperatives (Narain 2009, 11):

- One: to share the global commons equitably, because we know that co-operation is not possible without justice.
- Two: to create conditions so that the world, particularly the energy deprived world, can make the transition to a low carbon economy.

A Cosmopolitan Moment and Methodological Cosmopolitanism

Ulrich Beck (2009) in his most recent work entitled *World at Risk* proposes that we are in a "cosmopolitan moment" that emerges through the increased awareness of the dynamics of world risk society. Through this process "all people have become the immediate neighbors of all others, and thus share the world with non-excludable others, whether they like it or not, or want to recognize it, or not" (Beck 2009, 56). He states that cosmopolitanism is necessary since it "makes the inclusion of others a reality and/or its maxim." He, like Fraser, comments on the difference between cosmopolitanism as 'reality' and 'maxim.' He explains that in the normative sense (of 'maxim'), cosmopolitanism means *recognition* of the plight of the other. Narain (2009, 11) states explicitly that the challenge for the world is to "find ways of *learning from* the environmentalism of the poor, so that we can all share a common future" (my emphasis). Beck (2009, 56) proposes that the traumatic experiences of the "enforced community of global risks that threaten everyone's existence" holds the potential to be more inclusive, thus potentially allowing for subalterns to speak, meet with the more powerful, and deliberate the terms for resolving climate injustice globally. He states however, that:

> It would be too easy to assume that globality could spontaneously generate a shared global or planetary consciousness. The everyday experience of globality does not take the form of a love affair of everyone with everyone else. It develops in the perceived emergency of global threats to civilized action—in the networks, financial flows or natural crises mediated by information technology. . . . In other words it is the *reflexivity of world risk society* that founds the reciprocal relation between publicity and globality. With the constructed and accepted planetary definition of threats, a joint space of responsibility and action is

being created across all national boundaries and divisions, a space that *potentially* (though not necessarily) founds political action between strangers in an analogous way to the national space (Beck 2009, 182).

The move from defining commonly agreed upon threats to defining action, is notoriously difficult, as shown by the lack of success of international and transnational regimes in proposing clear cut actions for dealing with climate change and climate injustices to date. Glasser (2007) comments too that there is a need for social learning processes to address this gap—a gap that he sees between our stated desire for a more sustainable world and our everyday actions and policies. Beck proposes that addressing this gap *is indeed possible,* since the action that needs to develop emerges from what he refers to as our "enforced community of global risks" (2009, 56), but this will require addressing and raising awareness of the dominance of an "ego-istic selectivity with which the West responds to the world risk society." Such ego-istic selectivity is evident in the dominance of mitigation technologies and individuation of the problem to individual behavior in climate change response programs originating in the West (seen in the dominance of education programs focussing on the facts of climate change, the development of efficient technologies, and calls for individuals to switch off the lights, and reduce personal footprints—traces of which are eveident in the SADC REEP climate change educational context—see Table 4.1). These responses are ego-istic because rapid adaptation practices and societal and structural change responses are most necessary in those countries and local contexts most at risk from the impacts of climate change. Mitigation is a secondary issue in these contexts.

To address the problem of ego-istic responses to climate change, Beck proposes a move beyond approaches that try to solve world risk problems through methodological nationalism or contextualism only. Methodological nationalism, he argues, is a form of contextualism where global risks are not seen to exist, and risks and hazards are defined, distributed, and individualized in contextual, national, and local terms. This methodological problem is one of the root causes of ego-istic responses to climate change. It fails to take the wider climate justice issues into account. In the SADC REEP educational context, such ego-istic responses may also become more prominent with an emergent focus on localized adaptation practices and contextualized social learning. Ego-istic responses tend towards favoring situated learning approaches and constructivist learning since they focus on climate change risks and their resolution at a local or contextually significant level (i.e., the individual smallholder farm level, or on country-specific issues for example). The unfortunate outcome of ego-istic responses to climate change is exacerbation of the incommensurability that Fraser refers to in her analysis. Educationally, ego-istic responses would result in a narrowing of the possibilities for participation in reflexive justice dialogs, and a narrower range of options for thinking about, practicing, and contributing to the fundamentally transformative climate change responses asked for by Narain (2009).

Since it *is* also necessary to respond to climate change risks as they manifest in diverse ways in different parts of the world, there is a need for contextualism in educational work. It would be difficult to work out and learn appropriate alternatives and solutions to immediate and complex risks (i.e., to build capabilities and skills for risk negotiation in the everyday) without such an orientation, thus situated social learning is *necessary*. However, it would seem necessary to ensure that social learning processes pay adequate attention to the wider nature and scope of climate injustice *in addition to* understanding climate change risks and injustices at a local/contextualized level. To do this, Beck proposes "methodological cosmopolitanism," which involves developing a better understanding of transnational actor networks and how they contribute to the defining and distribution of risks such as climate change. It also focuses attention on social vulnerabilities and inclusion, and global reflexivity of global risks, and the need for ongoing deliberation and reflexivity in resolving these risks. Importantly, he does not propose a normative or political cosmopolitanism or a world-historical subject of cosmopolitanization. Beck's *methodological cosmopolitanism* introduces reflexivity and justice-centered responses to global risks such as climate change, since it forces everyone to recognize that *anyone* can be affected by these risks in principle, even if these effects are different in different places. It also begins to redraw the boundaries of old conflicts between North and South, center and periphery, which contribute to the incommensurability that Fraser mentions.

For educators it means researching climate change justice questions from the perspective that everyone shares the risk. It requires the development of a *transborder framework* for climate change education that is inclusive of social vulnerability, and global reflexivity where new proposals can emerge. Social learning methodologies in this sense need to include engaged *global and transboundary dialogs* to extend and provide wider meaning to locally constituted sustainability practices such as greening activities, nutrition practices, or carbon sequestration at a local level.

Beck (2009) states that coming to terms with global problems such as climate change through methodological cosmopolitanism necessitates participation in *global efforts* to address the problems. Through this methodological focus, Beck's work provides a vital reflexivity between the local and global, necessary for Fraser's reflexive justice to have meaning, and for a program like the SADC REEP to adequately engage with the nature and scope of climate change issues, particularly their relationship to social justice aspects of the issue. Beck (2009, 181) states that such 'globality' is not without conflicts or tensions, since contradictions such as when "industrialised countries call for the protection of important global resources, such as rainforests, while simultaneously reserving the lion's share of energy resources for themselves" exist. While these conflicts and contradictions are likely to continue to exist (as also pointed out by Fraser 2008; Worth

2009; Bond 2009), Beck points out that they also perform an integrative function, since they make it clearer that global solutions will need to be found, not through war, but through creating new global institutions and rules, and new frameworks for justice which can accommodate the transnational world risk society that has been created out of the twentieth century development frames, as also proposed by Narain (2009). Wals (2007) points to the value of conflict and contradiction in learning towards sustainability, as do other sociocultural learning theorists (Engeström 2005). Fraser (2008) argues that in seeking reflexive justice, it is important to draw constant attention to *mis-framings* of justice questions, one of which may be an overemphasis on contextualization of environmental education and adaptation responses, or localizing social learning processes *without* giving equal and adequate attention to their global nature, content, and origin.

SCENARIO 2: BROADENING THE SOCIAL LEARNING RESPONSE IN CLIMATE CHANGE EDUCATION

This chapter argues that climate change educational programs that are oriented towards wider societal transformation and emancipation would need to bring into focus the reflexive justice thesis of Fraser (2008), Narain's (2009) point that the world needs to *learn from* the environmentalism of the poor, and Beck's (2009) methodological cosmopolitanism. These methodological strategies provide a wider vantage point on climate change education than some discourses of social learning (interpreted through methodological nationalism or contextualism). Such methodological orientations, when used as *complementary to,* and *extending of* situated social learning processes, promise more for developing the sophistication, confidence, and will to power that are necessary for *speaking with authority* (Fraser 2008) in a context of severe uncertainty, risk, state-based hegemonies, and *efficiency* discourse responses (driven by market *habitus*) that exclude sufficiency and equity discourses. Such an approach should also allow transboundary dialogs and wider sharing of experiences that are emerging from what Narain refers to as an "environmentalism of the poor," a process that will hopefully help to counter ego-istic responses to climate change.

As outlined in the review of the SADC REEP's emerging climate change education work, there are many interesting dimensions to the climate change education program. Important to the wellbeing of people in the southern African region is the emphasis on adaptation practices and situated social learning processes that model, demonstrate, and encourage the emergence of new (alternative) practices that are likely to be more sustainable and that will extend and enhance socioecological resilience. This work is currently in its infancy. This is not surprising since the dominance of particular forms of ego-istic responses to climate change have led to adaptation suffering from what Schipper and Burton (2009) refer to as "benign neglect." From the low number and somewhat dated nature of the knowledge resources available on climate change issues and adaptation practices, it would seem that SADC itself has also neglected this work. Unlike in the cases of mitigation science and technology, there is much "uncertainty about adaptation to climate change, what the concept means or should mean, how adaptation can be effectively introduced, facilitated and managed, for whom, when and where" (Schipper and Burton 2009, ix).

This small scale review (see Table 4.1) also identifies that there appears to be little methodological or resource-based guidance on how to approach and respond to climate injustices in educational work at the SADC REEP. This chapter has tried to provide additional conceptual and critical resources for engaging with climate injustice in education, to inform the climate change education program at the SADC REEP. It argues that there is a need to upgrade and update the scope and nature of the climate change knowledge resources available at the SADC REEP, and that the SADC REEP could look into ways of strengthening a transboundary reflexive justice framework for climate change education that can *complement* and *extend* the contextual educational framework that has been established already. This will be necessary to avoid a localization of climate change responses, and to avoid possible conservativism and ego-istic orientations to climate change that can result from situated social learning processes and contextualized responses. If this were to happen, it would be paradoxical, since current efforts focusing on climate change education in the program are clearly increasingly oriented towards social transformation and addressing the problems caused by global climate injustices through an emphasis on adaptation practices.

Within this scenario, twenty to thirty years from now, one might find the farmers' associations that the SADC REEP worked with practicing a range of adaptation strategies that have ensured continued food security for their families. The extension officers supporting the farmers are successful in countering the unsustainable side effects of the new green revolution that has been brought into Africa through large-scale development initiatives that were co-conceptualized and implemented with African farmers. These development initiatives (for a change) are located in local experience, culture, and practices, as well as the best available scientific and technological knowledge. They are funded by international climate justice funds that allow the farmers to create new 'hybrid' solutions and sustainable farming practices that are successfully combating the worst uncertainty patterns experienced as a result of climate changes. All this is based on a set of strong advocacy and networking activities that the SADC REEP is coordinating through their international links and through various knowledge-sharing practices in which they are demonstrating to the world how farming practices in Africa and elsewhere are based on principles of sufficiency and sustainability. Farmers and extension services from other parts of the developed world are coming over to Africa *to learn from people in Africa how to farm more sustainably in the face of climate uncertainties and changes.* Curriculum policies in schools, universities, and colleges have also been transformed to incorporate this new knowledge of climate justice, sustainability, and rapid and flexible strategies for adaptation to climate change. The SADC REEP has an active network of national, regional, and international curriculum policy makers engaged in evaluating the outcomes of this work, and sharing it with the rest of the world to extend the attention given to climate justice; and schools everywhere are busy teaching children how to participate in sustainability practices that address climate change impacts at a local level, but also on a global level. People everywhere are learning to act locally, but also to act globally at the same time. Reflexive climate justice is being practiced on a daily basis by people in different parts of the world, and even though they don't know each other personally, they are aware that they are all neighbors in the fact of ongoing climate change–related risks.

CONCLUSION

In the final analysis, this chapter proposes social learning responses to climate injustice that embrace an emancipatory globally reflexive conception. It shows too that this needs to be consciously developed and resourced, a framework for climate change education that is *simultaneously critically transnational* and *globally reflexive* while also supporting *situated social learning processes* that are contextually located and oriented towards agency, capability, and risk negotiation in the everyday. Since it is located in one of the regions most at risk from climate change impacts, southern African environmental educators are in a strong position to share knowledge of how justice-oriented approaches to environmentalism can help inform sufficiency and equity solutions to climate change, to counter the hegemony of efficiency discourses, and to allow others around the world to learn what other solutions might be possible. The time for enabling climate justice is short, but the potential for creativity, reflexive justice, and socially inclusive orientations to climate change responses are wide open.

REFERENCES

Africa Geographic. 2007. *Africa Geographic. Special Report: Our overheating planet.* August. www.africageographic.com (accessed October 20, 2008).
Beck, U. 1992. *Risk society: Towards a new modernity.* London: SAGE.
———. 1999. *World risk society.* Cambridge: Polity.
———. 2009. *World at risk.* Cambridge: Polity.
Benhabib, S. 2006. *Another Cosmopolitanism.* Oxford: Oxford University Press.
Bond, P. 2009. A timely death? *New Internationalist: Climate Justice Taking the Power Back* 419: 14–15.
Chivers, D. 2009. Just or bust: Can the Copenhagen talks deliver climate justice? *New Internationalist: Climate Justice Taking the Power Back* 419: 19–23.

Edwards, A. 2005. Cultural historical activity theory and learning: A relational turn. Keynote address at Teaching and Learning Research Project Annual Conference, November 2005, at University of Warwick.

Engeström, Y. 2005. "Non scolae sed vitae discimus": Toward overcoming the encapsulation of school learning. In *An Introduction to Vygotsky*. 2nd ed., ed. H. Daniels, 157–76. London: Routledge.

Fraser, N. 2008. *Scales of justice: Re-imagining political space in a globalising world*. Cambridge: Polity.

Glasser, H. 2007. Minding the gap: The role of social learning in linking our stated desire for a more sustainable world to our everyday actions and policies. In *Social learning towards a sustainable world: Perspectives, principles, and praxis*, ed. A. Wals, 35–62. Wageningen, The Netherlands: Wageningen Academic.

Havnevik, K. 2007. The relationship between inequality and climate change. *News from the Nordic Africa Institute* 3: 8–11.

Hoffmann, M. 2007. The day after tomorrow: Making progress on climate change. *Radical Philosophy* 143: 2–7.

Intergovernmental Panel on Climate Change (IPCC). 2007. *Fourth Assessment Report*. Cambridge: Cambridge University Press.

Lotz-Sisitka, H. 2008. *A summary of the research component in the SADC Regional Environmental Education Programme 2004–2008*. Grahamstown, South Africa: Rhodes University Environmental Education Unit.

Lotz-Sisitka, H., L. Olvitt, M. Gumede, and T. Pesanayi. 2006. *Report 3: Education for sustainable development practice in southern Africa*. Howick, South Africa: SADC REEP.

Narain, S. 2009. A million mutinies. *New Internationalist: Climate Justice Taking the Power Back* 419: 10–11.

O'Donoghue, R. 2007. Environment and sustainability education in a changing South Africa: A critical historical analysis of outline schemes for defining and guiding learning interactions. *Southern African Journal of Environmental Education* 24: 141–57.

O'Donoghue, R. and H. Fox. 2009. *Have you sequestrated your carbon? A Share-Net resource book*. Howick, South Africa: Share-Net.

Republic of South Africa (RSA). 2007. *South African environment outlook report*. Pretoria: Department of Environmental Affairs and Tourism.

Schipper, L. E. F. and I. Burton, eds. 2009. *The Earthscan reader on adaptation to climate change*. London: Earthscan.

Southern African Development Community Regional Environmental Education Programme (SADC REEP). 2002. *Project Document*. Howick, South Africa: SADC REEP.

———. 2009. *The Southern African Development Community Environmental Education Programme*. Program flyer. Howick, South Africa: SADC REEP.

Taylor, J. 2009. Notes on sustainability practices at the SADC REEP. E-mail communications, February 20, 2009.

United Nations Education, Science and Cultural Organisation (UNESCO). 2004. *Education for all. The quality imperative: EFA global monitoring report*. Paris: UNESCO.

———. 2009. *Education for sustainable development and climate change. Policy dialogue 4: ESD and sustainable development policy*. Paris: UNESCO.

United Nations Environment Programme (UNEP). 2006. *Africa environment outlook: Our environment our wealth*. London: Earthprint.

Wals, A., ed. 2007. *Social learning towards a sustainable world: Perspectives, principles, and praxis*. Wageningen, The Netherlands: Wageningen Academic.

Worth, J. 2009. Power politics. *New Internationalist: Climate Justice Taking the Power Back* 419: 4–7.

5 The Environment, Climate Change, Ecological Sustainability, and Antiracist Education

George J. Sefa Dei

INTRODUCTION

Climate change is not independent of a body and its politics. The study of climate change is interlinked with questions of power, social difference, equity, and justice. If we do not address the problem of climate change there will be no justice or equity for humanity to speak about. This paper enthuses the possibilities for transformation by applying a critical antiracist lens to education for environmental sustainability. The paper engages both the 'social' and 'physical' and conceptualizes 'environment' broadly to include the sociocultural and the natural physical realms. Such conceptualization is reasonable given that in times of actual and looming climatic change/environmental crisis what is needed is an antiracist interrogation of the image of environment as a 'management' issue. For example, as pointed out later, we must trouble the particular conceptions of 'Earth/planet's peoples' as representing different ethnicities and cultures conflated into a singular humanity. Such representations, however well-intentioned, mask the implications of how we speak of the 'global' devoid of power, complicity, and responsibility. Such liberal representations claim notions of the 'self' and 'other' and the constitutive sociocultural identities as essential to understanding of human communities and our relationships to Earth's vulnerability, given current human ecological arrangements. We need to show how race/racism informs the production, distribution, and reception of representations of the social and natural environments at particular moments and the concomitant distribution of power and resources. In other words, we must understand what particular power relations and distribution and allocation of material and nonmaterial resources influence the way the environment is understood and related to cultures and social groups in society.

Questions of environment, health, peace, justice, and fairness are oftentimes linked. The problem with environmental education (and some other educations) has been the failure to link them—or, at least, a failure to link them other than at a rhetorical level (see Greig et al. 1987; Orr 1992). The health of the local community depends on the viability of the environment. Current concerns about climatic change are matters of global community survival. The environment is an all encompassing dimension of human life and survival. But, while climatic change is of major issue it is just an aspect of the big story of the health of the global environment. Global trends on the world's (physical) environments affect us in different ways (see Alston and Brown 1993; Bullard 1993). Consequently, environmental change is a matter of grave human concern. So how can we better understand our relationships to environments and develop a critical consciousness informed by key questions of power, social difference, equity, and justice? Human relationships with our environments must be looked at holistically as including the body, mind, spirit, Earth, and social, economic, political, and cultural justice.

There is no question that we are all implicated in sustaining the Earth's social and natural resources. Before we come to terms with our collective responsibilities it is equally important that we examine how we are differently complicit in the degradation of the Earth's resources through consumer and corporate greed. A small fraction of our humanity, the rich and powerful, consume a disproportionate share of the globe's resources. We need from that fraction in particular a change in lifestyle patterns. The other day I came across a saying that I found very profound. It was an exhortation for us to "live simply so that others may simply live"!

There are some key questions to keep in mind as we navigate ways to address the challenges of the climatic/environmental changes we are witnessing: What does antiracist education bring to the table in terms of an awareness of the differential power relations and allocation of material and social resources? How can antiracist education allow for multiple voices in a truly transformative environmental practice and education? How can racialized narratives (of popular environmental discourses) be confronted with countering antiracist strategies that help reorganize knowledge and discourse, and also transform actual (environmental) practice to promote sustainability? How can we collectively create a political, cultural-ideological frame of thought and social action, linking notions of environment, culture, development, and education? How do we relate questions of history, identity, freedom, and liberation to the political goal of achieving the sovereignty and dignity of all peoples? Posing these questions is a starting point for a good discussion of racism, environmental sustainability, and climate change. Environmental education is about the possibilities of critical, anticolonial education that can help learners and local communities subvert the 'colonial/imperial order' that scripts human groups and our relations with environments. The environment is not something to be controlled or owned. The environment is about living in peace and harmony with nature in order to ensure the sustainability of resources for future use.

Environmental education, as a process of teaching and learning about the nature, society, and culture interface, calls for engaging the local/social environments and multiple cultural knowings in diverse contexts. Environmental education must be geared to social and community advancement. Such education seeks to promote the satisfaction of local needs and aspirations through social equity and justice, and a respect for the fundamental freedoms and rights of all peoples. 'Education' here is understood as the varied strategies, ways, and options through which we come to know, understand, live, and act responsibly in a complex social world and strive for collective wellbeing. Environmental education hinges on such questions as the purpose, objective, and goals of learning about the society, culture, and nature interface, as well as the intellectual and political ends of scholarship and activism. Environmental education works best with an appreciation of local peoples' knowledge and how such knowing contains seeds of creativity and resourcefulness critical to self- and collective actualization. An important goal of environmental education is to address the unequal [power] relations between the 'official' (sometimes commercialized) knowledge and local/Indigenous/community knowledges. Local cultural resource knowledges point to the ways in which communities come to share power and resources in order to address questions of difference and inequity. When tapped as part of formal educational processes, local knowings offer some of the possibilities of social transformation. Consequently, as one of its foremost objectives, environmental education ensures that the pursuit of 'development' extends beyond material, economic, technological, and material constraints and possibilities to affirm the spiritual, emotional, and social-cultural dimensions of human existence.

Sustainability as a critical discursive approach to safeguarding the environment ought to be pursued through a human-centred prism in three interrelated aspects. First, the need to reclaim and affirm past, present, and future intellectual traditions, knowledge, and contributions to global history as a necessary exercise in decolonizing environmental knowledge. Second, the need to reflect on the present and to theorize the environment beyond any restrictive boundaries and confines. For example, environmental stress is more than climatic change. It involves particular social and political arrangements that come to the fore as we negotiate the society, nature, and culture interface. The environment matters politically, culturally, spiritually, and economically. To understand the nature of the problems and challenges facing humanity, we need to bring a critical reflection about what constitutes a collective consciousness of the environment. Third, we must ensure that local and marginalized/oppressed peoples set the terms of the 'development agenda' that implicates the knowledge and use of local environments in ways that allow peoples to contest and project their general wellbeing into the future. For example, we cannot continue to have multinational corporations, big governments, and a Western corporate global hegemonic structure defining the rules and terms of engaging local environments.

When these concerns are taken up in the pursuit of environmental education, it necessitates a shift away from a 'politics of negotiation' to a 'politics of transformation.' By this strategy I make reference to the politics whereby oppressed peoples freely rethink ways of addressing the challenges confronting their daily survival. This is only possible if we heal ourselves spiritually, mentally, and materially. This also calls for an affirmation of the sense of community, power sharing, social responsibility, and spiritual reembodiment. It requires an engagement with a new anticolonial project, which allows today's learners to define an agenda for environmental education recognizing local freedoms and the shared sense of community, belonging, and responsibility.

In the midst of the contingencies of globalism, corporate international capitalism, imperialism, and racism, a critical environmental education can hold some of the possibilities for social transformation. The promotion of environmental education must involve the lived experiences and knowings of multiple subjects and agents as they relate daily with their environments. Environmental education must promote a reading of humanness that allows all learners (as subjects) to connect to their past, heritage. and histories, not only in the geophysical sense, but also in the spatial and temporal sense of a shared collective existence. Obviously we cannot achieve this goal without a spiritual connection that reinvigorates and affirms our complex and multiple identities. The critical relevance of environmental education is also to give a voice to oral histories and cultures. All learners must be able to draw from their histories, cultures and traditions to be self-reflective of their experiences and be better prepared to face the enormity of global challenges, such as climate change. (See also Smith 1999; McGovern 1999; Semali and Kincheloe 1999.)

Development, climate change, and the environment are interlinked. An antiracist prism can contribute to a decolonizing project whereby 'development' is not about what people lack, or what they are expected to become, but instead, what people do for themselves. This shifts 'development' from a horse race into caring for the environment. Indigenous epistemologies, emerging from the colonized, can challenge the institutionalized narrative that upholds a hierarchical ordering of knowings about environments. Antiracist and anticolonial thinking allows us to work with a politics of identity affirmation, while bringing to the fore questions for discussion about the interconnections of power, difference, and resistance. School curricula cannot afford to mis-educate about the environment and the responsibilities of sustainability. A critical antiracist education has to be steeped in local communities' experiences, and be able to point to the political, social, historical, and spiritual connections of global humanity.

THE PERSONAL, POLITICAL, AND INTELLECTUAL PROJECTS

As a recently enstooled traditional Chief in Ghana, writing this paper could not have come at a more opportune time for me. Traditional community Elders have one major responsibility: working with their local communities to maintain and uphold the sanctity and sacredness of the natural environment. Throughout many local communities traditional Elders are leading unheralded attempts and initiatives to arrest the deteriorating conditions of local environments. Not surprisingly I was personally saddened upon hearing the news in November 2008 (later officially denied of course) that the Ghanaian state was about to convert a national forest reserve in the capital city, Accra, into a commercial center. The Achimota Forest is the only surviving green belt in the Accra metropolis and save for the fact that this story of land sale for commercial purpose was a hoax public reaction would have been consequential. The idea that a national forest reserve was to be sold to an entrepreneur to build a shopping mall was preposterous. At the time the Office of the President came out quickly to quell the rumours! The forest located in the Accra-Tema Motorway extension spans 900 acres, and has been gazetted as a Forest Reserve since 1930.

To reiterate, this chapter is informed by a host of political and intellectual questions; for example, how can we better understand our relationships to environments and climate change and develop a critical consciousness of sustainability that is informed by key questions of power, social difference, equity, and justice? (See Greig et al. 1987.) How do local cultural resource knowledges impact on ways of rebuilding environmentally and ecologically sound and sustainable communities? This speaks to the necessity of bringing the social, spiritual, cultural questions front and center to our discussions on the environment, climate change, and sustainability. The 'socialization of environment' must be approached through a critical focus on social justice and equity issues and we ask: what are the social determinants of sustainability? The question is informed by a theorization of the environment and climate change beyond the physical/natural confines given the inseparability of nature, culture, and society. In framing the social in discussions of the environment, climate change, and sustainability my concern is not simply with the sociocultural contexts and determinants of the environment, but also how we address the social wellbeing of the planet. An important objective is to make the linkage between the current agendas for international development, equity issues, climate change, and global environments.

In what follows I will briefly operationalize the notion of sustainability and race/antiracism and the environment discourse. This discussion is then followed by an equally brief historical trajectory highlighting how sustainability relates to my teaching, research, writing, and community politics in the three areas of: (a) development in the African context; (b) race and antiracism discourse and practice; and (c) Indigeneity/Indigenous knowledge and multiple ways of knowledge production. Such an approach is pertinent in bringing grounded knowledge to the understanding of environment, climate change, antiracism education, and sustainability. With this in mind, I address Gyau-Boake's (2001) concerns with the Akosombo Hydroelectric Project and the microclimatic changes experienced by the local peoples. A key question here is: what does it mean to teach and research on sustainability? A particular focus is the broad philosophical and theoretical implications of 'education for sustainability.' That is, the ways particular forms of education help sustain our environment by promoting a healthy planet when due attention is paid to equity and social justice issues. I conclude by discussing the urgent need of addressing climate change and sustainability through antiracism education, the case of Akosombo, and the missing dialog.

TOWARDS A WORKING DEFINITION OF SUSTAINABILITY

In linking debates about deforestation, climatic change, pollution, lack of biodiversity, and development issues, Jabareen (2008, 187) notes that today environmental discourse has been "globalized" and has "transcended national boundaries". It is asserted that there is no single definition of sustainability and there is the "fluid paradox of sustainability" (Jabareen 2008, 188). Since the influential 1980 World Conservation Strategy (Selby 2006) the idea of 'sustainability' has become a very contested, value-laden term. (See also World Commission on Environment and Development, WCED [1988].) Voinov (2008) argues that sustainability is to be understood in a given political context with particular goals, priorities, and vested interests (488). Sustainability is about relations between humans and the use of resources. Sustainability also refers to the stability of social and natural environments. Sustainability seeks to achieve a 'balance' and 'integrity' in use and renewal of social and natural resources and an understanding of the relational dimensions of society, culture, and nature. Sustainability speaks to particular social-political arrangements. In contesting sustainability, Voinov (2008, 497) further notes, 'regional sustainability' may be at odds with 'global sustainability.' For the West, where the issue of conservation is now at the top of the 'development agenda,' sustainability may be an enticing discussion. However, in the South, where sustainability has always been part of local cultural knowledge and social practice, there is nothing new. Hence, cultural and political dimensions are integral parts of discussing sustainability.

In proposing a working definition for sustainability, I would maintain that sustainability is the capacity of keeping a state of balance across society, culture, and nature interrelations in the long run, so as to ensure effective use and regeneration of social and physical resources. As Osorios et al. (2005) put it, an Indigenous conception of sustainability speaks of:

> . . . a particular cosmovision of Indigenous peoples who understand nature as a whole, as life itself. Therefore nature cannot be instrumentalized on the grounds of further material gains. Nature is mediated by ethic principles that are grounded, simultaneously, in cultural values built along centuries of harmonic coexistence with and within nature. (504)

Given the limits of sustainability in the exponential growth of capital, these authors (Osorios et al. 2005, 506) ask: how do we "generate new processes, alternative pathways that let us face our social, economic, political, environmental and cultural reality from a common point of view according to our historical cultural background?" It is thus instructive for Nath (2008, 471) to equally opine: does the Earth have a limited 'self-regenerative capacity'? Furthermore, according to Osorios et al. (2005, 507) sustainability has "Normative" (i.e., what it should be) and "Positive [Instrumental]" (what it actually is) dimensions. Jabareen (2008) also notes that sustainability can no longer be an "environmental logo." We must look at sustainability from the point of view of society and power relations rather than the environment. Sustainability raises questions about rights, ethics, and moral values; for example, human conduct as 'good' and 'evil,' the tensions between the 'domination of nature' and the 'intrinsic rights of nature.'

CONCEPTUALIZING ANTIRACISM, INDIGENEITY, AND CLIMATE CHANGE: THE CASE OF THE AKOSOMBO DAM AND THE VOLTA LAKE PROJECT

Antiracism is a discursive and political practice addressing the myriad forms of racism and their intersections with other forms of oppression. Antiracism argues that in order to understand the true effects of race, we must acknowledge the way in which race intersects with other forms of difference: gender, class, sexuality, ability, language, and religion. However, antiracism primarily addresses the systemic and institutional dimensions of racism and draws attention less to overt racist acts, lodged in individual actions, practices, and beliefs. Within society structural racism revolves around certain ontological, epistemological, and axiological foundations. For example, at the ontological level, our institutions are seen as fair, value free, and objective. At the epistemological level, it is argued that by working with 'merit,' 'excellence,' and a 'prism of thinking in hierarchies,' we (as workers) can arrive at the ontological foundation of fairness, objectivity, and value-free society. Axiologically, it is insisted that treating everybody equally (as in social justice for all) and discounting the qualitative value of justice is the most appropriate thing to do. Equal opportunity and color blindness are much heralded, despite the fact that these notions complicate racism by masking its real material and political effects and consequences. Critical antiracism challenges White power and its rationality for dominance. It also resists racist and colonial privileges. The academic and political project of critical antiracism is to uncover how Western civilization scripts local communities through the fabrication of Whiteness and the policing of its corollary racial boundaries (see also Fanon 1963; Scheurich and Young 1999; Cesaire 1972; Memmi 1991).

In connecting race, antiracism, climate change, and environmental sustainability I am inviting readers to 'read race into environment and environmentalism.' Environmental racism is conceptualized in two important dimensions: first, the ways privileged segments of society consume and overexploit global and common resources to the detriment of environmental sustainability for all; and, second, understanding the objectives and goals of teaching and learning about environmental and ecological sustainability. Dominant society believes in an achievement ideology and in the hardworking individual so deserving of the benefits and luxuries they enjoy in society. Some may offer a hypocritical reading that turns the gaze away from questioning the impact of our activities on the environment, and instead blame the less fortunate as neither ecologically prudent nor engaging in activities that sustain the environment. We also see this when the West blames the South for environmentally-unsound practices and activities. Yet the West turns a 'blind eye' on their activities (e.g., pollution, fuel use, and consumption). We also have contemporary examples of multilateral corporations and big companies dumping industrial waste with environmentally hazardous consequences on the surroundings of the least-advantaged segments of our populations (see also Collin and Harris 1993; Proceedings to the First National People of Color Environmental Leadership Summit 1991).

Let us take, for example, the microclimatic changes that occurred since the construction of the Akosombo dam, the seemingly innocent hydroelectric project that allowed for the anthropogenic formation of the Volta Lake in Ghana. To sum up Gyau-Boakye's (2001) arguments, Akosombo was supposed to help usher Ghana into the industrialized realm of 'modernity.' In doing so, the project created jobs, such as "fishing, farming and transportation" (ibid. 18). But Gyau-Boakye also amplifies the problematic impact of this anthropogenic project. He notes that when the project was envisioned in the interest of 'development' and industrialization, what was lacking was a thorough "environmental impact assessment" (ibid. 17), that the concern was consumed with the technical outcome of generating electricity for an aluminum-smelting company. More so, the Akosombo engendered physical, biological, and environmental problems. Gyau-Boakye writes about "reservoir induced seismicity (RIS)" (ibid. 20), and the "morphological change" (ibid. 20) whereby the Volta's entry into the sea shifted somewhat to the east. This shift has also been linked to "coastal erosion" (ibid. 21). More importantly, Gyau-Boakye reminds us that Akosombo has been "linked to the increase in endemic water borne disease, that of, 'bilharzias, malaria and onchocerciasis'" (ibid. 21,22,24) and that one of the biological impacts that materialized were the weeds that have come to hinder the flow of the lake, harming fishing as a way of life. He points to the fact that rainfall in the forest and savannah regions of West Africa had diminished, leading to drought, while at the same time there was high evaporation resulting from a one-degree-Celsius rise in temperature. Also, Gyau-Boakye describes the local peoples and their stories concerning the change in wind speed on and around the Volta Lake, due to "bare or vegetated land being transformed into a body of water" (21). He also points to the social problems emanating from the Akosombo, such as "resettlement and the loss of lands and cultural practices" (25); in particular, the emotional pain and, as Indigeneity tells us, the "spirit injury" incurred due to loss of ancestral practices. Indigeneity is conceptualized in this context as a knowledge consciousness that comes with long-term occupancy of a place. Indigeneity is engaged as the nexus of culture, society, and environment. Taken up discursively the matter of Indigenous spirituality then becomes a knowledge base [a cosmovision] grounded in understanding of local environments—environment as sacred and the sacredness of activity. A focus on the community context of practices that promote and enhance sustainability would reveal the sacredness of activity that is held among Indigenous peoples relating to their environments. Far from the critiques of romanticism and overmythicization, what is acknowledged is the tensions of 'tradition' and 'modernity.'

The important theoretical lessons here would include an acknowledgment of the dynamism of culture, place/land, identity, and knowledge in addressing climate change; the evocation of the environment/land in contestations over history, language, identity, and heritage. This implicates how we understand the Indigenous claims of land, identity, culture, and space. For example, when the material relations are presented as the sole or primary basis of identity construction and articulation it serves to displace notions of culture, place, land, spirituality, and race . . . which have long been central in the construction of traditional/Indigenous identities of many colonized peoples (see Kempf 2008).

There is also the issue of 'embodied knowing' when we speak about climate change, environment, and sustainability, that is, the environment as a site of knowledge rather than simply as a resource. This critical reading of 'embodiment' extends the focus on "the kinds of social and political relations that are established with bodies, minds, and senses" (Titchkosky 2007, 5). Hence 'embodiment' of the environment is "a socially mediated phenomenon." We need to understand knowledge of the environment as being embodied. This is so important in informing us that the "objective knowledge of the physical environment" is different from the experiencing of the physical environment 'in the flesh' (subjective knowledge about the environment). The objective analysis of the environment does not fully equip one to understand the actual experience of engaging the environment.

This is a form of knowledge that is embodied (i.e., embodied knowing). It is not rooted in biology but instead arises from social experiences of a subject/body in a context where such knowledge about the environment is salient (see also Howard 2008). How Indigenous peoples come to understand and work with "the entities of nature, plants, animals, stones, trees, mountains, rivers, lakes and a host of living entities . . . all constitute embodied relationships that must be respected" (see Cajete 2000, 78).

Indigenous conceptions of land, spirituality, and culture all acknowledge the sacred history, ancestral homelands, ceremonial cycles, and language (see Taiaiake and Corntassel 2005). Many of the communities work with a concept of "radical Indigenism" (Garroutte 2003, 144) using all available Indigenous talents to begin a process of decolonization and regeneration.

SUSTAINABILITY AND THE 'DEVELOPMENT' QUESTION

In my PhD dissertation titled *Adaptation and Environmental Stress in a Ghanaian Forest Community* I was interested in how local communities survived socioenvironmental and economic stressors (e.g., drought, returning migrants, and contraction in the national economy) (Dei 1986). My focus at the time was on local pragmatism (e.g., the judicious use of the local environment). The study pointed to the reliance on local creativity and resourcefulness. I also set out to document planting and harvesting knowledge of local agricultural systems and traditional pharmacology (Dei 1986, 1990, 1992).

The major theoretical lessons of development and ecological sustainability in my study emphasized the environment as a resource and knowledge base that stresses the nexus of 'environment, culture, nature and society' (see also Dei 1991, 1993; Warren et al. 1995). It was obvious that the physical environment/Earth has life of its own and the environment is a community good, i.e., the common property resource argument (Berkes 1989). There was also the need to reformulate the environmental degradation and economic growth debate. While environmental degradation has an effect on development it is such unchecked economic growth (as development) that causes the degradation in the first place. This theoretical approach was a debunking of the poverty-environmental degradation thesis. Poverty is not an independent variable. Likewise climate change is not independent of a body. Poverty is not a predator or a destroyer of the environment. Environmental degradation is caused more by the ambitions of economic growth rather than by poverty (Escobar 1995). The fact is that local peoples will not conserve resources (even if to their benefit) if they will not be enjoying the benefits of their restraints. What all this pointed to was the urgency of rethinking the idea of 'development' (see Munck and O'Hearn 1999). In his critique of the "myth of development" Tucker (1999) notes that "the model of development now widely pursued is part of the problem rather than the solution" and that this Western ideology of development "distorts our imagination, limits our vision, [obscuring] us to the alternatives that human ingenuity is capable of imagining and implementing" (1). In order to imagine new possibilities Southern peoples must deconstruct the myth of development using local and Indigenous cultural knowings through an anticolonial prism.

Munck (1999) rightly argues that much of the ongoing intellectual discussion on 'development' is caught up in the dominant paradigms of Western thinking. Alternative visions and counter theoretical perspectives of development even struggle to disentangle themselves from the dominance of the Eurocentric paradigm. Eurocentrism is intrinsic in the ways we think, conceptualize, and organize knowledge about the environment. The whole debate about environmental sustainability has been shaped by the cultural forces and the political agendas of the West (Sardar 1999, 52). The question of who controls the 'environmental' discourse, and why, is significant especially when we begin to interrogate issues of power and resistance. Sardar (1999) in his excellent critique of 'development' has rightly argued that the issue at hand is one of having the definitional power of the West in control of the discourse about 'development.' The same can be applied to conventional ways of talking about 'the environment.'

The tendency to blame local peoples' cultural practices, customs, and traditional values and attitudes and beliefs as obstacles to environmental sustainability is rife. There are only sparing attempts made to critically interrogate Euro/Westcentric, capital-driven development strategies. As Gueye (1995) long ago noted, in the context of Africa, there is a specific cultural understanding in communities that is " . . . centred around a particular conception of the world which assigns the human being a specific role, around a certain representation of time and space which structures mentalities and behaviours . . . " relating to the their environments (11). In effect, development is what people already know and do using their creativity and resourcefulness, not what they lack or are expected to become.

CONCLUSION: RETHINKING SUSTAINABILITY AND CLIMATE CHANGE: THE QUEST FOR ANTIRACIST EDUCATION AND HEALTHY COMMUNITIES — AKOSOMBO AND THE MISSING DIALOG

In conclusion, I would bring an extended reading to the discussion of antiracist education for environmental sustainability and climate change, one that centers the social self and the collective. There is a social denial of the fact that we have an 'unhealthy problem' of 'divided communities' within nations, one that goes to the root of structural and systemic inequities. There are systemic inequities that foster youth alienation, poverty, and class disenfranchisement, loss of hope and dream. There is also violence associated with some current practices of 'development' at the local levels, which is compounded by the fact that we usually adopt a limited and narrow definition of environmental violence. In other words, it is not just the material and physical violence from current development/environmental practices but also a spiritual, emotional, psychological violence ('spirit injury') that afflicts most of our communities and their peoples. Part of this violence also results from blurring the conceptual distinction between asking communities to take responsibility for their problems and pathologizing families/communities for the problems they have to contend with.

The 'environmental crisis' is a making of modernity (see Reznai-Moghaddam and Karami 2008, 408). It is about a particular ordering of society where asymmetrical power relations among groups, individuals, and communities position and implicate us differently in acknowledging responsibilities and complicities. In the name of progress, humanity has experienced a movement of modernity in which industrialization has become the order of the day, whereby particular geographies, particular geobodies have come to know themselves as less than humans. Under the tropes and guise of 'development' we have seen the blatant disregard for safeguarding environments for human use. It is thus important an antiracist education for sustainability should move beyond the duality of the physical and social to a more holistic conceptualization of the environment—natural, physical, and social space; for example, the environment as spiritual, as a cultural resource knowledge. Case in point: the Akosombo and Indigenous communities. Akosombo, a modernization project developed on the "firm ground" discursive (Selby 2006), saw more than fishing, farming, and transportation spaces opening up to the local peoples. Instead, there was a displacement of the human, loss of cultural practices, ailments, and microclimatic alterations which resulted in increased earthquakes, rise in temperature, less rainfall leading to drought, change in wind speed in the local regions, and environmental health hazards such as malaria and bilharzias (see Gyau-Boakye 2001). We must consider Indigenous spirituality (cosmovision) and embodied knowledges as starting points, as operating central to the present day development/environmental practice.

Antiracist education for sustainability should distinguish between the environment as a resource (rights) perspective and as a site of knowledge production. There are theoretical and methodological limitations when the focus is one and not the other, that of the propertization of the environment, that the environment is simply a material resource. The issue of embodied knowing/embodiment, i.e., the environment is understood fundamentally as part of the living and the connections of the physical and metaphysical realms of existence.

Akosombo pushes us to ask new and critical questions concerning sustainability and climate change; sustainability as currently conceptualized is a 'privileged discourse'—sustainability for whom, why, and how? Who benefits from the focus of sustainable use of resources, and who has been practicing sustainability for years without acknowledgment? No doubt within the Akosombo project there were struggles of human/Indigenous existence; this calls for understanding the saliency of the land (e.g., struggles over language, culture, identity, politics). It means bringing to the fore pointed questions of responsibility and accountability that implicate all of us but have differential and uneven responsibilities (e.g., the notion of 'live simply so that others may simply live').

An antiracist education for sustainability must engage the possibilities of an interdisciplinary approach to studying the environment. This calls for addressing the "impermeability and inaccessibility of scientific language . . . a distinctive mark of positivist thought, still in vogue in the Western world, and through which linguistic barriers keep different disciplines isolated" (Osorios et al. 2005, 515). The study of the environment, climate change, and the question of sustainability must enhance rather than serve to undermine the agency of local peoples. Local subjects' voice and agency are critical for environmental education. There are also practical components in terms of teaching as praxis. Research into the environment and sustainability must be informed by everyday experiences of what people actually know, do not know, about the environment.

Antiracist education for sustainability should be more than educating to sustain human lives. Such education should also work with the uncertainty of knowing as healthy ways of knowing, (e.g., the ontological primacy of interpretations)—how we make sense of our worlds: how does interpretation influence experience and what are the respective resources for knowing about the environment? On the power of local knowledge we should inquire as to the extent to which in rebuilding communities after natural and human-made disasters such as the Tsunami that occurred in Asia and Hurricanes Katrina and Wilma, and civil wars, how much we use local knowledges in such undertakings? In other words, what can the Akosombo Hydroelectric Project learn from Indigenous models of environmental sustainability? How can the subjects of our work become the theorists informing the Akosombo project? In effect antiracist education for 'environmental sustainability' must take into account local understandings of the workings of culture, society, and nature. The emotional and spiritual wellbeing of the individual and the social group is the bedrock of any environmental sustainability process. Environmental sustainability is socialization of knowledge. The notion of environmental sustainability must be invoked in the name of the common good. Environmental sustainability must seek to appropriate long-standing traditions of mutuality and sustainability to meet local needs and aspirations. Antiracist education for environmental sustainability means matching the rights of individuals to membership of particular groups and at the same time locating corresponding social responsibilities. Antiracism education regarding the environment is taking action to protect children, elderly, women, and the wellbeing of marginalized communities through a healthy environment. The experience of racism itself is known to cause mental illness, hypertension, and other emotional distress among the oppressed (see also McDonald 2002; Lerner 2004; Shrader-Frechette 2002; Stein 2004). Aside from the spiritual and emotional wounding of racism, we also know that when it comes to issues of health, climate change, and the environment we can link the effects of pesticides, chemicals, and pollutants that surround human communities to deteriorating health conditions (see Thompson 2008). The poor and disadvantaged segments of our communities can least afford the rising cost of health care. From Akosombo to Western societies, we know that childhood illness can be attributed to industrial chemicals and environmental pollutants (see also Thompson 2008). On Native reserves in Aboriginal communities we learn about children suffering from respiratory illnesses, leukemia, brain tumors, and other illnesses. Antiracist education can be critical in informing local citizens, businesses, and decision makers about environmental issues, about climate change, about why we need to be concerned about greening cities, protecting people, and saving species from extinction. The destruction of the ecosystem impacts on human survival, and the most vulnerable groups any time the environment is threatened are the poor, disenfranchised, and marginalized segments of our communities. Antiracist activism is working for change, making links between environmental pollution and health problems and drawing the connection between toxins and serious human health concerns. In building sustainable healthy communities we ought to deal with power inequities that exist in our communities, we ought to create environmentally sound societies, we ought to challenge the power issues and confront the problematic separation of the material and the nonmaterial. What is called for is a 'project of restoration;' that is restoring hope, visions, and the dreams of communities.

SNAPSHOT

The year 2050 is upon us; the quest for modernity pushes on relentlessly, capitalist development bulges at the foreground. Indigenous knowledge is but a mere whisper, drought persists, earthquakes are closer to us than before, malaria is more of a household name by now, decrease in rainfall, increase in temperature are all expected, this all quietly shaped through the landscape of the Akosombo. Conversations concerning climate change continue to be voiced, more of them coming from the West, those from the local have reorganized the voice of the West as theirs. The urge to talk about the Akosombo and the environment, through fables, folklore, and Indigenous proverbs has fallen by the wayside. Alienated from within, through history and time are embodied knowings and Indigenous cultural practices. Colonial governance is organized through the World Bank and IMF, and their 'Poverty Alleviation Strategies' continue to inhabit Indigenous populations. The global outcry vociferously begs for the green solution. Climate change dialog, independent of 'bodies,' is more and more in our everyday conversation. The void continues to be filled through the problematization of capitalist greed and multinational corporations. Elsewhere, another scenario resides in the hallways of academe where the relevance of antiracism education is perpetually challenged and resisted through historic-scientific-technological epistemologies. The echo is that racism is in the past, was of another time, that, we need to move on. We are in a post-racial era. The question of African 'development' and sustainability continues to place Africa as the source of their problems; Africans must take responsibility for their problems. Antihistorical readings of Africa and Indigenous peoples' lives through mass media. Sustainability marches on towards the Western front. Loss of humanism comes alive through statistics. Charts, tables all animated, all geared to race Africa to the First World Pole. Deforestation resides well. Voices of Indigenous spirituality are all but viewed as superstition. Talk of greening is with us everywhere. The 'us' and 'everywhere' being privileged, media-popularized geographies. Where then is the Indigenous voice? And sustainability for whom? Omnipresent race is absent from boardrooms. Environmentalists push on for greenery. Local peoples cling to their way of life. Oral histories breathe life into the environment, but to no avail. Much has been lost. What now? Where do we go from here? Local communities point to spirituality and oral cultures for help. The global community point to development, and dance to the tune of capitalist modernity. The shout is progress. The hope of humanism is all but lost. What we have are our dreams. Long gone is the hope of antiracism. Long gone is the climate.

REFERENCES

Alston, D. and N. Brown. 1993. Global threats to people of colour. In *Confronting environmental racism: Voices from the grassroots*, ed. R. D. Bullard, 179–94. Boston: South End Press.

Berkes, F. 1989. *Common property resource: Ecology and community-based sustainable development*. New York: Belhaven Press.

Bullard, R. D., ed. 1993. *Confronting environmental racism: Voices from the grassroots*. Boston: South End Press.

Cajete, G. 2000. *Native science: Natural laws of interdependence*. Santa Fe, NM: Clear Light Publishers.

Cesaire, A. 1972. *Discourse on colonialism*. New York: Monthly Review Press.

Collin, R. W. and W. Harris. 1993. Race and waste in two Virginia communities. In *Confronting environmental racism: Voices from the grassroots*, ed. R. D. Bullard, 93–106. Boston: South End Press.

Dei, G. J. S. 1986. *Adaptation and environmental stress in a Ghanaian forest community*. [Unpublished PhD diessertation, University of Toronto.]

———. Indigenous knowledge and economic production: The food crop cultivation, preservation, and storage methods of a West African rural community. *Ecology of Food and Nutrition* 24 (1): 1–20.

———. 1991. The dietary habits of a Ghanaian farming community. *Ecology of Food and Nutrition* 25 (1): 29–49.

———. 1992. *Hardships and survival in rural West Africa*. Dakar: CODESRIA. [Also published/reproduced in French in 1993].

———. 1993. Sustainable development: Revisiting some theoretical and methodological issues. *Africa Development* 18 (2): 97–110.

Escobar, A. 1995. *Encountering development. The making and unmaking of the Third World*. Princeton, NJ: Princeton University Press.

Fanon, F. 1963. *The wretched of the earth*. New York: Grove Press.

Garroutte, E. M. 2003. *Real Indians: Identity and survival of Native Americans*. Berkeley, CA: University of California Press.

Greig, S., G. Pike, and D. Selby. 1987. *Earthrights: Education as if the planet really mattered*. London: Kogan Page/World Wildlife Fund.

Gueye, S. P. 1995. Science, culture, and development in Africa. *CODESRIA Bulletin*, 7–12.

Gyau-Boakye, P. 2001. Environmental impacts of the Akosombo Dam and the effects of climate change on the lake levels. *Environment, Development and Sustainability* 3: 17–29.

Howard, P. 2008. Black gaze on the white ally: A critical race Africology of white antiracism. [Draft chapter of unpublished PhD dissertation, Department of Sociology and Equity Studies, Ontario Institute for Studies in Education of the University of Toronto.]

Jabareen, Y. 2008. A new conceptual framework for sustainable development. *Environment, Development and Sustainability* 10: 179–92.

Kempf, A. 2008. *The anti-colonial discursive framework: Toward an application to the Cuban educational context.* [PhD comprehensive paper, Department of Sociology and Equity Studies, Ontario Institute for Studies in Education of the University of Toronto.]

Lerner, S. 2004. *Diamond: A struggle for environmental justice in Louisiana's chemical corridor.* Boston, MA: MIT Press.

McDonald, D. A. 2002. *Environmental justice in South Africa.* Athens, OH: Ohio University Press; Cape Town: University of Cape Town Press.

McGovern, S. 1999. *Education, modern development, and indigenous knowledge: An analysis of academic production.* New York and London: Garland.

Memmi, A. 1991. *The colonizer and the colonized.* Boston: Beacon Press.

Munck, R. 1999. Deconstructing development discourses: Of impasses, alternatives and politics. In *Critical development theory: Contributions to the new paradigm*, ed. R. Munck and D. O'Hearn, 195–209. London: Zed Books.

Munck, R. and D. O'Hearn. 1999. *Critical development theory: Contributions to the new paradigm.* London: Zed Books.

Nath, B. 2008. A heuristic for setting effective standards to ensure global environmental sustainability. *Environment, Development and Sustainability* 10: 471–86.

Orr, D. 1992. *Ecological literacy: Education and the transition to postmodern world.* New York: State University of New York Press.

Osorios, L., M. Lobato, and A. Del Castillo. 2005. Debates on sustainable development: Towards a holistic view of reality. *Environment, Development and Sustainability* 7: 501–18.

Proceedings of the First National People of Color Environmental Leadership Summit. 1991. October 24–27, Washington, DC.

Reznai-Moghaddam, K. and E. Karami. 2008. A multiple criteria evaluation of sustainable agricultural development models using AHP. *Environment, Development and Sustainability* 10: 407–26.

Sardar, Z. 1999. Development and the location of Eurocentrism. In *Critical development theory: Contributions to the new paradigm*, ed. R. Munck and D. O'Hearn, 44–61. London: Zed Books.

Scheurich, J. and M. Young. 1999. Coloring epistemologies. *Educational Researcher* 26 (4): 4–16.

Selby, D. 2006. The firm and shaky ground of education for sustainable development. *Journal of Geography in Higher Education* 30 (2): 351–65.

Semali, M. and L. J. Kincheloe, eds. 1999. *What is indigenous knowledge?: Voices from the academy.* New York and London: Falmer Press, Taylor & Francis.

Shrader-Frechette, K. S. 2002. *Environmental justice: Creating equity, reclaiming democracy.* Oxford, New York: Oxford University Press

Smith, L. 1999. *Decolonizing methodologies: Research and indigenous peoples.* London and Dunedin, New Zealand: Zed Books and University of Otago Press.

Stein, R. 2004. *New perspectives on environmental justice: Gender, sexuality, and activism.* New Brunswick, N.J.: Rutgers University Press.

Taiaiake, A. and J. Corntassel. 2005. Being indigenous: Resurgences against contemporary colonialism. *Government and Opposition* 40 (4): 597–614.

Thompson, S. 2008. Environmental justice in education: Drinking deeply from the well of sustainability. In *Green frontiers: Environmental educators dancing away from mechanism*, ed. J. Gray-Donald and D. Selby, 36–58. Rotterdam: Sense.

Titchkosky, T. 2007. Pausing at the intersections of difference. In *Exploring gender in Canada: A Multi-dimensional approach*, eds. B. Matthews and L. Beaman, 135–37. Toronto: Pearson Prentice Hall.

Tucker, V. 1999. The myth of development: A critique of Eurocentric discourse. In *Critical development theory: Contributions to the new paradigm*, ed. R. Munck and D. O'Hearn, 1–26. London: Zed Books.

Voinov, A. 2008. Understanding and communicating sustainability: Global versus regional perspectives. *Environment Development and Sustainability* 10: 487–501.

Warren, D. M., L. J. Slikkerveer, and D. Brokensha. 1995. *The cultural dimension of development: Indigenous knowledge systems.* London: Intermediate Technology Publications.

World Commission on Environment and Development (WCED). 1988. *Our common future.* London: Department of the Environment [on behalf of the United Kingdom Government].

6 Learning in Emergencies
Defense of Humanity for a Livable World

Fumiyo Kagawa

INTRODUCTION

'Emergency' or 'crisis' is a socially constructed and contested notion. However, there is increasing recognition that human beings and planet Earth are facing interlocking global challenges, or emergencies: increasing numbers of localized wars, a growing volume of refugees and internally displaced populations (IDPs), serious environmental degradation, loss of biological diversity, deep levels of poverty, and the HIV/AIDS pandemic. Challenges we face will be further deepened, not least because of still unknown consequences of global warming and impending abrupt climate change. We are at the crossroads:

> At the state for the 21st Century, we too are confronted with the fierce urgency of a crisis that links today and tomorrow. The crisis is climate change. It is still a preventable crisis but only just. (UNDP 2007, 1)

Severe impacts of climate change have already been felt in various parts of the world and it is anticipated that climate change will cause great stress on both the environment and society. There is an increasing awareness that climate change will severely impact or undo international development efforts, such as the United Nations Millennium Development Goals (MDGs) (UNDP 2007; UNICEF 2008; UNICEF UK 2008). MDGs as agreed at the 2000 United Nations Millennium Summit aim at achieving the following by 2015: eradication of extreme poverty and hunger (Goal 1); achieving universal primary education (Goal 2); promoting gender equality and empowering women (Goal 3); reducing child mortality (Goal 4); improving maternal health (Goal 5); combating HIV/AIDS, malaria, and other diseases (Goal 6); ensuring environmental sustainability (Goal 7); developing a global partnership for development (Goal 8) (UN 2006). For instance, UNICEF UK (2008) has examined MDGs with particular reference to children's survival, health, and education around the world. Climate change's adverse impacts on agricultural productivity, which contributes a large part of GDP in most developing countries, would lead to further poverty and hunger in such countries thereby undermining Goal 1. When malnutrition is already a significant cause of infant and child mortality, declining food productivity and water stress will also impact on the survival and health of children thereby putting the achievement of Goal 4 into reverse. Malaria, a cause of death for 800,000 children every year, will expand to previously less affected or nonaffected areas and would cause epidemics. Not only children, but also pregnant women will particularly be at high risk of malaria infection (so reversing the realization of Goals 4, 5, and 6). Climate change will make it more difficult for children to attend schools, especially for girls, who will be under more pressure than boys to support their families' survival needs. Subsequently, achieving universal primary education (Goal 2) and promoting gender equality and empowering women (Goal 3) become more difficult as the climate change scenario unfolds. It is also important to highlight that the impacts of climate change will not be evenly experienced. The economically poorer countries, children in the world, places, and population sectors that have done relatively little to trigger climate change would be hit by global warming "first and worst" (UNICEF UK 2008, 35).

In examining the links between climate change, peace, and war in detail, Smith and Vivekananda (2007) point out the additional pressure on societies that are already economically, socially, and politically fragile and that would have "a low capacity to adapt to climate change and face a high risk of violent conflict" (3). "Many of the world's poorest countries and communities," they state,

> face a double-headed problem: that of climate change and violent conflict. . . . There are 46 countries—home to 2.7 billion people—in which the effects of climate change interacting with economic, social and political problems will create a high risk of violent conflict. (3)

While Smith and Vivekananda's analysis primarily focuses on the sociopolitical and economic, UNEP (2009) particularly highlights the link between violent conflicts in relation to the exploitation of natural resources and environmental stress. It states:

> . . . the potential consequences of climate change for water availability, food security, prevalence of disease, coastal boundaries, and population distribution may aggravate existing tensions and generate new conflicts. (5)

In the face of the potential increase in number, intensity, and duration of humanitarian crises, arguably directly or indirectly triggered by climate change, this chapter reflects upon current and potential contributions of the field of emergency education (i.e., educational responses to humanitarian crisis situations).

CURRENT STATUS OF THE FIELD OF
EMERGENCY EDUCATION

Emergency education is an emerging field of study (Seitz 2004; Sinclair 2002; Sommers 2002). Because of an increase in organized violence in the form of war, civil strife, armed conflict, and political oppression since the end of the Cold War in 1989 as well as the increase in the number of natural disasters in the 1990s, education in emergency and crisis situations has become a major concern for the international community (Retamal and Aedo-Richmond 1998; Seitz 2004; UNESCO 2000). Education is increasingly recognized as the 'fourth pillar' of humanitarian aid in such crises, along with food and water, shelter, and health care (Machel 2001). There is a growing awareness that education must be a priority of emergency assistance. Emergency education is a broad field covering diverse formal, nonformal, and informal educational initiatives to support populations severely affected by conflict, disaster, or instability. Specific target groups include refugees, IDPs, nonmigrants, returnees, different gender and age groups (e.g., early childhood, primary school age, secondary school age, youth, adolescent, and adult), special need groups (e.g., child soldiers, excombatants, disabled children, orphans), and minorities (Sinclair 2001; UNESCO 2006).

There are a few recurring rationales within emergency education discourse, rationales that are not always mutually exclusive. A predominant view regards education as a human right, claiming that access to education is an inalienable right for all children despite their circumstances. The 1989 United Nations Convention on the Rights of the Child sets out the overall framework for any discussion of education and emergencies/crises. As a subset of this view, some emphasize a view that education is humanitarian protection. The proponents of this view think attending school is a measure of protection since it physically keeps the children away from risks, and through education, children can also learn the skills and knowledge to cope with increased risks, which in turn allows them to protect themselves (Nicolai 2003; Nicolai and Triplehorn 2003). There are others who consider that education should play a critical role in meeting psychosocial needs for children and adolescents affected by crises to heal their wounds and break the cycle of violence. Addressing psychosocial needs is seen as a prerequisite to learning anything after traumatic experiences (Aguilar and Retamal 2009; Machel 2001). In addition, there is the view that education is a long-term social investment for development (e.g., World Bank 2005), and that education in emergency contexts plays an important role in addressing an interconnected development and security agenda especially after the 9/11 terrorist attacks (e.g., World Bank 2003).

Looking at the field through two main categories of response to climate change, mitigation and adaptation, it is clear that the field has by and large lacked direct mitigation interests and efforts in reducing CO_2 emissions, while the field can be construed as contributing to climate change adaptation—reducing unavoidable impacts of climate change—through its engagement in humanitarian crisis contexts, arguably directly or indirectly triggered by climate change. Although a conscious link between climate change and the field has not been made among most of its proponents, there are growing concerns among agencies working in the field. They anticipate the enormous future challenges that humanitarian and development work would face because of the instability posed by climate change. Save the Children (2007) points out the potential limitations of traditional coping mechanisms and past experience as a guide to the future, and calls for attention to be paid so that "the education available for children is suitable to the changing environment" (10). It also urges the need for "new ways of working, imaginative solutions . . . an active engagement of children and their communities" (13). What follows examines conflicting potentials within the field in dealing with the serious impacts of climate change.

Boundaries of Emergencies

According to proponents of emergency education, 'emergencies' often fall into two broad categories: natural disasters (e.g., hurricane/typhoon, earthquake, flood, and drought) and human-made crises (e.g., war, internal conflict, and genocide) (Obura 2003; Pigozzi 1999; UNESCO 2006). In addition to these, some highlight silent/chronic emergencies such as persistent poverty, growing numbers of street children, and the HIV/AIDS pandemic (Pigozzi 1999; Williams 2006). Others highlight the abrupt nature of the events, stating that an emergency is "a condition which arises suddenly, and the capacity to cope is suddenly and unexpectedly overwhelmed by events" (Hernes 2002, 2). There are also those who emphasize a longer time span of emergency by regarding it as "encompassing not only the first days or months after an event, but also the effort to deal with the ongoing effects of the crisis, and reconstruction" (Nicolai and Tripleholn 2003, 2). Complex emergencies are "situations that are 'man-made' (sic) and are often caused by conflict or civil unrest, which may be compounded by a natural disaster" (Inter-Agency Network for Education in Emergencies 2004, 7).

It is important to note that the identification of these different types of emergencies does not necessarily mean that the field responds to them equally. There are conflicting views on the current emphasis of the field. For instance, Chand et al. (2003) observe that emergencies caused by natural disasters remain rather a secondary concern compared to conflict-triggered emergency education. On the other hand, Save the Children (2006) notes that educational responses to emergencies caused by conflict are still lacking, although education became a familiar aspect of humanitarian operations in recent natural disasters, such as the 2004 Asian Tsunami, and earthquakes in Iran, India, Pakistan, and Indonesia. There exist clear political bias and differentiated interests with regard to intervention among the donor countries and agencies (Renner and Chafe 2006; Williams 2006). In addition, considering the fact that chronic and multiple crises and vulnerabilities are the everyday reality of the majority of people especially in the global South (Gonzáles-Gaudiano 2005; O'Donoghue and Lotz-Sisitka 2006; Williams 2006), it is problematic that 'emergencies' or 'crises' to which the field currently responds are narrowly focused and selective in their nature. Dominant emergency education does not deal with silent and chronic emergencies at the implementation level "except in so far as they occur during situations arising from armed conflict or natural disaster" (Sinclair 2002, 23), and the field lacks critical examination of the root causes of emergencies. What becomes critical is to address the issue of power: Who defines emergencies? Whose emergencies are prioritized and why? What motivates the external educational interventions in response to emergencies and why? How can participation of those who are affected by emergencies be better promoted in the context where the external international organizations often possess the financial and material resources? (Kagawa 2007, 2009) The selective nature of constructing emergency in the field also suggests an underlying assumption that a particular crisis can be dealt with in isolation and that 'human-made' emergencies are independent of nature.

Another issue regarding the current conceptualization of emergency in the field is its time-bounded phases/stages. For instance, UNESCO (2006) suggests the following categories: acute outset, protracted emergencies, return and integration, and early reconstruction. Such linearly conceived divisions are perpetuated by organizational mandates and funding structures linked to the different emergency phrases, types, and particular population groups. This, in turn, becomes an underlying cause of short-sighted and compartmentalized educational responses to emergencies (Sinclair 2002; Smith and Vaux 2003; Sommers 2002; UNESCO 2006).

A linear temporal conceptualization of emergencies is clearly predicated upon the assumption that emergencies start and end, normality is recovered, and development continues. When the consequences of climate change deepen and become prolonged, emergencies would not be contained within specific space and time. How would the priorities of an international agency's interventions be justified when there are continuous demands to respond to unending emergencies in various locations? The higher end of climate change projections suggested by a number of scientists make such concerns more than a mere brainstorming exercise. This issue will be revisited later in the chapter.

Emergency as a Window of Opportunity

Although the tendency to stick with the familiar old educational system from before the crisis erupted is normally strong, crises are often seen as

windows of opportunity for rethinking the old system and introducing radical innovations within the system as a whole (Williams 2006). Since people
tend to accept the notion of a crisis/emergency as a window of opportunity
rather uncritically, it is important to unpack what the opportunity means
especially from the dominant business-as-usual perspective. Klein (2007)
appraises how a moment of mass suffering and collective trauma triggered
by a flood, a hurricane, a war, a terrorist attack, a debt crisis has been easily manipulated by an ideology of "disaster capitalism" over the past thirty
years. By disaster capitalism she means "orchestrated raids on the public
sphere in the wake of catastrophic events, combined with the treatment of
disasters as exciting market opportunities" (6). While the whole population
is deeply shocked, paralyzed, and disoriented by a great rupture, disaster
opportunists take advantage of "a blank slate" in order to implement quick
and radical social and economic reforms in favor of the global free market
economy. She points out that crises "are, in a way, democracy-free zones—
gaps in politics as usual when the need for consent and consensus do not
seem to apply" (140).

When considering characteristics of learning which best address the
needs of those who are affected by various forms of calamity, it is vital to
be aware of the wider context of the intensifying force and speed of disaster
capitalism, which commodifies and exploits local human and natural 'capi-
tals' in a shocked society. Developing local capacities to resist predatory
global market forces individually and collectively is one of the important
elements to be included in the theory and practice of emergency education.
Yet, this is a pertinent lacuna in a field which does not problematize the
notion of development (Kagawa 2005, 2009), and education tends to play
an instrumental role in perpetuating mainstream development ideology
and practice. For instance, according to the World Bank (2005):

> much of post conflict reconstruction in education involves familiar de-
> velopment activities, the "usual business," of developing and reforming
> the education system—yet it is usual business in very unusual circum-
> stances. (26)

Here the 'usual business' referred to is work on all the fronts of familiar educational change tasks (e.g., national education policy development,
curriculum development, teacher training, material and textbook development, development of an educational administration system). In spite of
the caveat that 'usual business' does not mean 'business as usual,' such an
interest is deeply embedded. It states that one of the roles of education in
post-conflict situations is to get "the country onto an accelerated development path" (World Bank 2005, 27).

Figure 6.1. depicts the dynamic relationship between unsustainability,
emergencies, modes of development, education/learning, and a resilient
social-ecological system (sustainability). The main thesis is that as far as

educational responses to emergencies are predicated on and lean towards dominant neoliberal development and globalization (i.e., the left side of sequences in Figure 6.1), they are status quo confirmative by perpetuating the existing exploitative and unequal systems.

In tackling the challenges of climate change through both mitigation and adaptation efforts, it is vital to address the root cause of climate change. That is the dominant development discourse and practices upon which the economically and materially privileged minority of the populations and societies have predicated and projected their worldview for more than two centuries. Gardiner (2008) elaborates this point as follows:

> The source of climate change is located deep in the infrastructure of current human civilizations; hence, attempts to combat it may have substantial ramifications for human social life. (29)

He goes on to say that:

> . . . given that halting climate change will require deep cuts in projected global emissions over time, we can expect that such action will have profound effects on the basic economic organization of the developed countries and on the aspirations of the developing countries. (29–30)

However challenging, unless it is recognized that the climate crisis fundamentally stems from dependence on fossil fuel–based economy and unless the dominant assumption and practice of limitless economic and material growth in a finite Earth (at enormous cost of human and ecological wellbeing) is addressed, what is suggested results in pseudo-solutions or market solutions that only make the situation worse and induce yet further rounds of crisis. We need to go beyond a business-as-usual mentality and solutions (Shiva 2008). To sum up, in the words of Hossay (2006):

> The trouble is, increasingly, the only goals that matter are those defined by the market. Concerns over the health of the global ecosystem, justice, traditions, sacred beliefs, shared community, care and concern for fellow beings, are all left by the wayside. (120)

ENHANCING RESILIENCE

The right side of sequences in Figure 6.1 depicts alternatives to status quo confirmative responses to emergencies. To better cope with ongoing, unknown, yet inevitable consequences of climate change, the field needs to address dominant development discourse first and foremost. Moving away from hegemonic, narrow, market-driven development is critical. From an alternative perspective, development involves multidimensional processes and concerns, improvement of wellbeing, and quality of life to meet the needs of human and nonhuman nature (Haavelsrud 1996). It has to be "self-directed, self-regulated and self-organized evolution from within" (Shiva 2008). Development should also be redefined in a way that

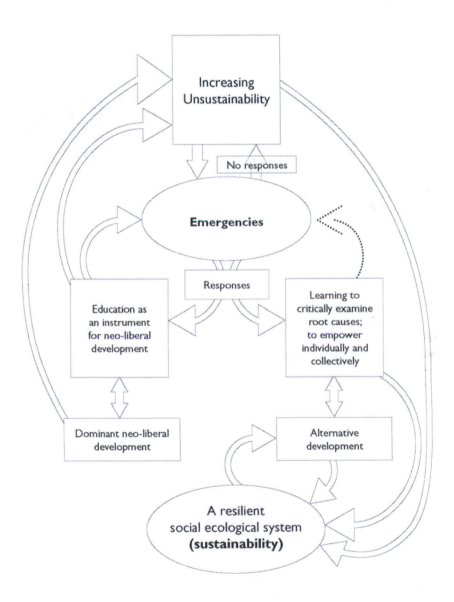

Figure 6.1   Dynamic responses to emergencies (inspired by Folke et al. 2008, 359).

enhances human and ecological resilience. According to Folke (2002), the concept of resilience is:

> the capacity to absorb sudden change, cope with uncertainty and surprises while maintaining desirable functions. Resilience provides the components for renewal and reorganization following change. (228)

In contrast to a vulnerable system which would break down faced with even small changes and reverberations, Folke points out that "in a resilient system, change has the potential to create opportunity for development, novelty and innovation" (228). Folke further states that "the degree to which the social-ecological system can build and increase the capacity for learning, adaptation and responding in a manner that does not constrain or erode future opportunities is a central aspect of resilience" (229).

In the field of emergency education, the concept of resilience has already been employed among those who are working on 'natural' disasters from the disaster risk reduction perspective. Among its proponents, disasters are distinguished from natural hazards (e.g., cyclone, earthquakes, floods) and disaster risk is considered to arise "when hazards interact with physical, social, economic and environmental vulnerabilities" (UN/ISDR 2005, 1). This idea is expressed in the following well-used equation: "Risk = Hazard x Vulnerability" (UN/ISDR 2004, 71). Disaster risk reduction has two key components: minimizing vulnerability and increasing resilience (physical, social, economic, environmental) and the preparing for disasters at multiple levels including within both formal and nonformal education (Save the Children 2008; Twigg 2007; Wisner 2006). The disaster risk reduction approach also emphasizes the importance of creating a 'disaster resilient community,' or 'sustainable community' which is characterized by mutual trust, respect for cultural diversity (including indigenous knowledge), and active community participation and self-reliance (UN/ISDR 2004; Twigg 2007). Rather than dealing with symptoms separately, it addresses underlying vulnerabilities comprehensively.

In exploring the pedagogical implications of the notion of resilience, Miller and Affolter's (2002) questions are helpful:

> . . . What are the deep sources of resilience that carried [those who have been affected] through the crisis? How can interventions respect their strength and enable communities—and children themselves—to act as the primary agents of educational reconstruction? (117)

In relation to Miller and Affolter's first question, diversity plays a key role in enhancing resilience. For instance, Suzuki and McConnell (1997) argue that biological, genetic, and human cultural diversities are indispensable to human and natural long-term survival in withstanding catastrophes. Using some examples of vulnerability in large-scale monocultural agriculture, fishery, and forestry to any perturbations such as diseases/pests and fire, they affirm that ecological diversities are vital for long-term survival of species.

They also point out the mutually reinforcing nature of ecological and cultural diversities. Diverse ecosystems give rise to diverse cultures and, in turn, diverse cultural traditions and knowledge help to conserve diversities of the planet. So protection of ecological diversities and diverse indigenous knowledge systems should go hand in hand. Extrapolating this to learning, diverse ways of knowing and expression, diverse speeds of learning, as well as diverse sources and forms of knowledge become important. The learners also need to gain a profound understanding of the interconnectedness and interdependence of ecological and cultural diversities (Kagawa 2009).

In response to Miller and Affolter's second question, it becomes vital to "reassemble the ontological fabric of a society" that is scattered by a crisis by reconstructing a devastated community culturally, physically, ontologically (Bush and Saltarelli 2000, 31), and environmentally. Such a view also gives a pedagogical rationale for employing various media with those affected by traumatic events to enable them to understand difficult experiences and express themselves. Aguilar and Retamal (2009) and UNICEF (2006) affirm that curricula purely focused on cognitive outputs are far from breaking a cycle of violence and healing the psychological scars of children and youth affected by crises. Employing various pedagogies for self-expression (e.g., music, dance, arts, sports, storytelling) is vital for "quality humanitarian education" (Aguilar and Retamal 2009).

In addition, and very importantly, narrative can play an especially critical role in resisting the strong influence of disaster capitalism that exploits the shocked society. Klein (2007) states:

> Any strategy based on exploiting the window of opportunity opened by a traumatic shock relies heavily on the element of surprise. A state of shock, by definition, is a moment when there is a gap between fast-moving events and the information that exists to explain them. . . . Without a story, we are . . . intensely vulnerable to those people who are ready to take advantage of the chaos for their own needs. As soon as we have a new narrative that offers a perspective on the shocking events, we become reoriented and the world begins to make sense once again. (458)

Individual and collective memories are, Klein claims, the "greatest shock absorber of all" (463). Echoing the importance of memories in post-crisis learning, Folke et al. (2008) refer to "creating social and institutional space or platforms for dialogue and innovation" as "key to stimulating learning and to resolving social uncertainties" (360).

BEYOND ADAPTATION

A number of commentators point out that we must stop atmospheric temperature rise at two degrees Celsius. Beyond that, major ecosystems begin to collapse and "climate change is out of our hands: it will accelerate without our help" (Monbiot 2007, xxi). The world of a five-degree-Celsius atmospheric temperature rise, which Lynas (2007) thinks might arrive in a few decades if urgent actions to radically reduce greenhouse emissions have not been taken, faces "a pace of warming much too rapid for substantial adaptation either by natural ecosystems or human civilization" (208). It is the world characterized by catastrophic mass extinctions, large-scale human displacements and starvation, and violent conflicts over land and food beyond our imagination. "Needless to say, the era of food aid and international assistance would be long gone" (ibid. 210). When runaway climate change effects are intensified through a positive feedback loop, the distinctiveness of the field of emergency education becomes default or all education becomes emergency education in that educational initiatives everywhere need to address abrupt and/or chronic crisis situations one way or another (Selby 2007).

Although no one can predict exact future scenarios in particular time and space, what is certain is that no one can escape from the serious consequences of climate change and "the world will experience significant and potentially highly dangerous changes in climate over the next few decades *no matter what we do now*" (italics in original; Walker and King 2008, 53). In considering the implications of increasingly dire scenarios of climate change, or creeping emergencies,[1] two key aspects of education are considered here drawing upon insights from the field of emergency education: roles of learning and organizing learning.

Roles of Learning

In the context of severe survival struggles, a fundamental question for educators is: how can we respond to adversities "without losing our humanity" (McIntosh 2008)? As human history shows, such contexts have often triggered a negative downward spiral of violence characterized by self-centered, short-sighted, and human-centered thinking and behaviors, a sense of insecurity, distrust, and hatred of 'others.' A recent example includes the 2005 Hurricane Katrina, which witnessed crime, host community hostility, or even racism to outsiders/evacuees (Cannovó 2008). In Darfur, although the causes of violent conflict are complex, the confluence of intensifying poverty, marginalization, and migration triggered by "climate variability, water scarcity and the steady loss of fertile land" can "create the conditions that make violence an attractive option for disempowered young men" (UNEP 2009, 9).

When coexistence—not only among different human groups but also between human beings and nonhuman animals and nonhuman nature—is under threat, to expand our sense of compassion to 'others' becomes critical. In the words of Opotow et al. (2005), this means extending "our scope of justice" that is "the psychological boundary within which concerns about fairness govern our conduct" (305). Those outside the boundary of justice are morally excluded and "beyond our concerns, and eligible for deprivation, exploitation, and other harms that might be ignored or condoned as normal, inevitable, and deserved" (ibid. 305). In the context of climate crisis, future generations, who have to deal with all the consequences of today, need to be within our scope of justice. All species should also be within the scope of justice if we want to sustain a livable planet. Pedagogically, expanding our boundaries of justice is not simply about obtaining technical skills and rational knowledge. Our own taken-for-granted perspectives need to be challenged and values such as compassion, empathy, appreciation, and frugality should be cultivated.

Second, kindling and sustaining a sense of hope should be a moral mission for educators, by helping learners overcome a sense of fatalism, cynicism, or passivity. Cultivating a sense of hope becomes particularly important especially when life will be full of incremental and slow decline in livelihood, ecosystem, health, economy, as well as the sudden onset of extreme weather-related disasters (IPCC 2007). Bartlett (2008) points out that "slowly debilitating conditions (along with repeated exposure to more extreme weather) may dull the awareness necessary for proactive responses" (89). For Klein (2007), a sense of hope is cultivated through collective actions and resultant sense of ownership over one's own life:

> The universal experience of living through a great shock is the feeling of being completely powerless: in the face of awesome forces, parents lose the ability to save their children, spouses are separated, homes—places of protection—become death traps. The best way to recover from helplessness turns out to be helping—having the right to be part of a communal recovery. . . . Radical only in their intense practicality, rooted in the communities where they live, these men and women see themselves as mere repair people, taking what's there and fixing it, reinforcing it, making it better and more equal. Most of all, they are building in resilience—for when the next shock hits. (465–66)

Those collective actions should also go hand in hand with addressing broader structural issues that trigger the crises. In this sense, crisis could open a window of opportunity to create a more just and equal society.

Third, meeting learners' psychosocial needs becomes vital when they face a frequent sense of loss on multiple fronts: loss of family and friends, loss of home, loss of place-based community, loss of social networks, loss of cultural diversity, loss of routines and jobs, loss of individual and collective identity, to name a few (Cannovó 2008). As mentioned earlier, various forms of self-expression can help but pedagogical approaches should focus strongly on meeting a human being's basic psychosocial needs (e.g., a sense of security, positive connection with other individuals and groups, positive self-awareness and identity, comprehension of reality). When these needs are satisfied, Miller and Affolter (2002) affirm, those affected by crises begin to regain a sense of wellbeing.

Organizing Learning

One of the possible consequences of the increase of both acute, chronic, as well as creeping emergencies in the future is the constant interruption of formal education provision in various parts of the world. The field of emergency education has witnessed that when formal education is interrupted, informal and nonformal education play a complementary role. Hence, it is very likely that in a climate-constrained future society, boundaries between formal and non/informal education will inevitably become permeable. In other words, the foci and leadership of learning need to become more flexible depending upon changing circumstances. Top-down, externally and expert-driven educational provisions and learning processes will simply become obsolete or dysfunctional, and more dispersed and horizontal forms of leadership and knowledge creation will become more imperative. Considering the fact that even now it is local people who need to deal with the consequences of emergencies long before and after the external interventions (Jabry 2002; Plan 2005), enhancing local leadership and ownership in learning is a viable option for the future.

Organizing learning in more flexible and nonhierarchical ways inevitably leads to the issue of knowledge creation. A review of the field of emergency education has revealed that in the contexts awash with a sense of urgency, delivering externally created knowledge (e.g., externally premade teaching/learning resources) and focusing on quantity, not quality, of learning have often been justified in the name of efficiency and urgency, at the expense of validating local needs and knowledge (Kagawa 2009). When crisis-affected people are not consulted about their "longer term needs," even those who "go into disaster situations with the best of intensions," Miller (2005) warns, "run the risk of seeking short-term fixes at the expense of doing longer-term permanent damage" (2).

What is needed is a sea change in the direction of flow of knowledge both at theoretical and at practical levels. Those who are engaged in education in emergency situations should be very mindful that "people in emergencies are their own experts in their experience" (John Barry,[2] personal communication). More emphasis should be given to building up knowledge emerging from crisis-affected local people especially in the global South, which is currently desperately needed in the field (Aguilar and Retamal 2009). New processes of engagement—where local knowledge and experiences are heeded and respected, and acted upon—should be actively promoted. Voices from the margins—children, women, the poor, those with disabilities, minorities—should be heard through culturally appropriate arenas and channels. In fact, to respond to climate chaos, Shiva (2008) believes that real solutions come from those who are marginalized, especially from the poor in the global South who know how to live lightly.

In creating a change in the flow of knowledge, those who are in the 'upper' hierarchy equipped with material privileges and affluences, who are primarily responsible for the climate crisis, must be more self-aware and open to learning. Questions raised by Chambers (1997, 100–01) become important reminders for educators positioned at the 'upper' end of hierarchy:

> All powerful uppers think they know
> What's right and real for those below
> At least each upper so believes
> But all are wrong; all power deceives . . .
> Whose knowledge counts?
> Whose values?
> Whose criteria and preferences?
> Whose appraisal, analysis and planning?
> Whose action?
> Whose monitoring and evaluation?
> Whose learning?
> Whose empowerment?
> Whose *reality* counts?
> 'Ours' or 'Theirs'? [italics in original]

He goes on to ask (101), "What can and should we, as uppers, do to make our realities count less, and the realities of lowers—the poor, weak and vulnerable—count more?"

CONCLUSION

In the face of the ever-increasing number and intensity of humanitarian crises, as well as the creeping emergencies posed by climate change, it becomes critical not only to respond but also to anticipate them by addressing underlying multiple vulnerabilities in a way that enhances social and ecological wellbeing. Reflecting upon the following words of McIntosh (2008) is apposite in concluding this chapter: "We live in strange times that can offer strange openings" (227). Climate change can bring an opening for human beings to collectively confront root causes of interconnected global injustice and inequalities; an opening for those who are in materially privileged positions to pursue right (and light) living on the finite earth; an opening for all of us to extend our moral and ethical horizons; an opening for us to be more humble and appreciative of life.

FUTURE SCENARIOS

To end this chapter, two imaginative scenarios for the year 2059 in Japan are depicted. Now the country is not free from acute and chronic emergency situations because of very unstable and often violent weather conditions, although it is fortunate enough not to have any armed conflicts.

Scenario 1

It has been more than a few decades since Japan was compelled to open its door to accommodate a flux of environmental refugees from the neighboring Asian and Small Island countries according to the 2019 UN Humanitarian Treaty. The number of those who have lost their homes and livable land because of floods caused by intense typhoons, rising sea levels, as well as storm surges in this region has already surpassed 2.5 million, a figure estimated by Walker and King (2008) as likely to be reached by 2050. Japanese land is indeed crowded, food and energy are constantly in scarcity, and medical services are overburdened and at the verge of collapse. The country's demographic compositions have significantly changed and it is now a multicultural society.

The country has just been hit by another massive typhoon. Like the previous one of a few weeks ago, this one has caused floods in the large coastal areas in the western part of the country. It is never easy for people and never the same even after experiencing quite a few typhoons. At the Takamatsu evacuation center located within a local elementary school building, there are a significant number of youth who are taking a very active role in helping others—the elderly, the sick, and the injured—by assisting medical staff in offering life-saving support and information. They are also taking a leadership role in organizing informal learning activities for younger children. In addition, they play an active role as interpreters among people from different countries of origin (the younger generations have become fluent in two, even three languages and can help older generations who can only speak their own native language). There are a number of circles where people share their own stories of survival. A retired teacher, 75-years-old, said, "Although life is very tough and I am very sad to see so much loss and suffering during a time like this, I still feel hopeful when I see those young people acting very courageously and generously for others. Maybe the educational transformation introduced a few decades ago—which emphasized the importance of coexistence, appreciation, and action, and, also, most importantly, learning to live with change and uncertainty has borne some fruit."

Scenario 2

The United Nations University in Tokyo is hosting a five-day international conference titled 'Climate Change, Emergency, and Learning.' It has now become common sense for people around the world to avoid flying as much as possible especially for attending a short conference like this, so the majority of the participants who live outside of Japan attend via videoconference or Internet facility. The electronically present speakers and presenters consist of the poor, the marginalized, the noneducated (in a formal sense) from the global South, who have been hit very hard by climate change impacts. They would never have been considered as speakers at such a prominent international conference a few decades ago. Educational experts and researchers from the North have finally learned to listen to and validate the everyday, but sacred, knowledge of the South.

NOTES

1. The term was coined by Professor David Selby during discussions with the author in June 2004.
2. Reader, Department of Politics, International Studies and Philosophy at Queen's University Belfast, Northern Ireland

REFERENCES

Aguilar, P. and G. Retamal. 2009. Protective environments and quality education in humanitarian contexts. *International Journal of Educational Development* 29: 3–16.

Bartlett, S. 2008. The implications of climate change for children in lower-income countries *Children, Youth and Environments* 18 (1): 71–98.

Bush, K. and D. Saltarelli. 2000. *The two faces of education in ethnic conflict: Towards a peace building education for children*. Florence: UNICEF Innocenti Research Centre.

Cannovó, P. 2008. In the wake of Katrina: Climate change and the coming crisis of displacement. In *Political theory and global climate change*, ed. S. Vanderheiden, 177–200. Cambridge, MA: MIT Press.

Chambers, R. 1997. *Whose reality counts? Putting the first last*. London: Intermediate Technology Publications.

Chand, V. S., S. Joshi, and R. Dabhi. 2003. Emergency education: The missing dimension in education policy. *Educational Research for Policy and Practice* 2: 223–35.

Folke, C. 2002. Social-ecological resilience and behavioural responses. In *Individual and structural determinants of environmental practice*, ed. A. Biel, B. Hansson, and M. Martenensson, 226–42. Aldershot, UK: Ashgate.

Folke, C., J. Colding, and F. Berkes. 2008. Synthesis: Building resilience and adaptive capacity in society in social-ecological systems In *Navigating social-ecological systems: Building resilience for complexity and change*, ed. F. Berkes, J. Colding, and C. Folke, 352–87. Cambridge: Cambridge University Press.

Gardiner, S. 2008. A perfect moral storm: Climate change, intergenerational ethics, and the problem of corruption. In *Political theory and global climate change*, ed. S. Vanderheiden, 25–42. Cambridge, MA: MIT Press.

González-Gaudiano, E. 2005. Education for sustainable development: Configuration and meaning. *Policy Future in Education* 3 (3): 243–50.

Haavelsrud, M. 1996. *Education in developments*. Tromsø: Arena.

Hernes, G. 2002. Responding to emergencies. *IIEP Newsletter*, July–September.

Hossay, P. 2006. *Unsustainable: A premier for global environmental and social justice*. London: Zed Books.

Inter-Agency Network for Education in Emergencies (INEE). 2004. *Minimum standards for education in emergencies, chronic crises and early reconstruction*. Paris: Inter-Agency Network on Education in Emergency (INEE).

Intergovermental Panel on Climate Change (IPCC). 2007. Summary for policymakers. In *Climate Change 2007: Impacts, adaptation and vulnerability*.

Contribution of Working Group II to the Fourth Assessment Report of the Intergovernmental Panel on Climate Change, ed. M. L. Parry, O. F. Canziani, J. P. Palutikof, P. J. van der Linden, and C. E. Hansen. Cambridge: Cambridge University Press.

Jabry, A. 2002. *Children in disasters: After the cameras have gone*. London: Plan UK.

Kagawa, F. 2005. Emergency education: A critical review of the field. *Comparative Education* 41 (4): 487–503.

———. 2007. Whose emergencies and who decides? Insights from emergency education for a more anticipatory education for sustainable development. *International Journal of Innovation and Sustainable Development* 2 (3/4): 395–413.

———. 2009. *Navigating holistic and sustainable learning: Challenges and opportunities in ongoing and anticipated emergencies*. [PhD dissertation, University of Plymouth.]

Klein, N. 2007. *The shock doctrine*. London: Penguin.

Lynas, M. 2007. *Six degrees: Our future on a hotter planet*. London: Harper Perennial.

Machel, G. 2001. *The impact of war on children*. Vancouver: UBC Press.

McIntosh, A. 2008. *Hell and high water: Climate change, hope and the human condition*. Edinburgh: Birlinn.

Miller, T. 2005. Forward. In *Children and the tsunami*, Plan, 2. Bangkok: Plan International Asia Regional Office.

Miller, V. and F. Affolter. 2002. *Helping children outgrow war, SD Technical Paper No. 116*. Washington, DC: USAID.

Monbiot, G. 2007. *Heat: How we can stop the planet burning*. London: Penguin Books.

Nicolai, S. 2003. *Education in emergencies: A tool kit for starting and managing education in emergencies*. London: Save the Children UK.

Nicolai, S. and C. Triplehorn. 2003. The role of education in protecting children in conflict. *Humanitarian Practice Network Paper 42*. London: Overseas Development Institute.

Obura, A. 2003. *Never again: Educational reconstruction in Rwanda*. Paris: UNESCO International Institute for Educational Planning.

O'Donoghue, R. and Lotz-Sisitka, H. 2006. Situated environmental learning in Southern Africa at the start of the UN Decade of Education for Sustainable Development. *Australian Journal of Environmental Education* 22 (1): 105–13.

Opotow, S., J. Gerson, and S. Woodsode. 2005. From moral exclusion to moral inclusion: Theory for teaching peace. *Theory into Practice* 44 (4): 303–18.

Pigozzi, M. 1999. *Education in emergencies and for reconstruction: A developmental approach*. New York: UNICEF.

Plan. 2005. *Children and the tsunami*. Bangkok: Plan International Asia Regional Office.

Renner, M. and Z. Chafe. 2006. Turning disasters into peacemaking opportunities. In *State of the world: Global security*, ed. Worldwatch Institute, 115–224. London: Earthscan.

Retamal, G. and R. Aedo-Richmond, eds. 1998. *Education as a humanitarian response*. London: Cassell.

Save the Children. 2006. *Rewrite the future: Education for children in conflict-affected countries*. London: International Save the Children Alliance.

———. 2007. *Legacy of disasters: The impact of climate change on children*. London: Save the Children UK.

———. 2008. *In the face of disaster: Children and climate change*. London: International Save the Children Alliance.

Seitz, K. 2004. *Education and conflict: The role of education in the creation, prevention and resolution of societal crises—consequences for development cooperation.* Federal Ministry for Economic Cooperation and Development. http://www.gtz.de.

Selby, D. 2007. As the heating happens: Education for sustainable development or education for sustainable contraction? *International Journal of Innovation and Sustainable Development* 2 (3/4): 249–67.

Shiva, V. 2008. *Soil not oil: Climate change, peak oil and food insecurity.* London: Zed Books.

Sinclair, M. 2001. Education in emergencies. In *Learning for a future: Refugee education in developing countries,* ed. J. Crisp, C. Talbot, and D. Cipollone, 1–83. Geneva: UNHCR.

Sinclair, M. 2002. *Planning education in and after emergencies.* Paris: UNESCO International Institute for Educational Planning.

Smith, A. and T. Vaux. 2003. *Education, conflict and international development.* London: Department for International Development.

Smith, D. and J. Vivekananda. 2007. *A climate of conflict: The links between climate change, peace and war.* London: International Alert.

Sommers, M. 2002. *Children, education and war: Reaching education for all (EFA) objectives in countries affected by conflict.* Washington, DC: World Bank.

Suzuki, D. and A. McConnell. 1997. *The sacred balance: Rediscovering our place in nature.* Vol. Vancouver. Vancouver: Greystone Books/Douglas & McIntyre.

Twigg, J. 2007. *Characteristics of a disaster-resilient community: A guidance note (Version 1 for field testing).* London: DFID Disaster Risk Reduction Interagency Coordination Group.

UN. 2006. *The Millennium Development Goals report.* New York: United Nations Department of Economic and Social Affairs.

UNDP. 2007. *Fighting climate change: Human solidarity in a divided world, Human Development Report 2007/8.* New York: Palgrave Macmilan.

UNEP. 2009. *From conflict to peace building: The role of national resources and the environment.* Nairobi: UNEP.

UNESCO. 2000. *The Dakar framework for action: Education for all: Meeting our collective commitment.* Paris: UNESCO.

———. 2006. *Guideline for planning education in emergencies and reconstruction.* Paris: UNESCO International Institute for Educational Planning.

UNICEF. 2006. *Education in emergencies: A resource tool kit.* Katmandu: UNICEF ROSA.

———. 2008. Climate change and children: A human security challenge. Florence: UNICEF Innocenti Research Centre.

UNICEF UK. 2008. *Our climate, our children, our responsibility: The implications of climate change for the world's children.* London: UNICEF UK.

UN/ISDR (International Strategy for Disaster Reduction). 2004. *Living with risk: A global review of disaster reduction initiatives.* Geneva: UN/ISDR.

———. 2005. *Hyogo Framework for Action 2005–2015: Building the resilience of nations and communities to disasters. Extract from the final report of the World Conference on Disaster Reduction.* Geneva: UN/ISDR. http://www.unisdr.org/eng/hfa/hfa.htm.

Walker, G. and D. King. 2008. *The hot topic: How to tackle global warming and still keep the lights on.* London: Bloomsbury.

Williams, P. 2006. *Achieving education for all: Good practice in crisis and post-conflict reconstruction.* London: Commonwealth Secretariat.

Wisner, B. 2006. *Let our children teach us!: A review of the role of education and knowledge in disaster risk reduction.* Bangalore: Books for Change.

World Bank. 2003. *'Mind the Gap': The World Bank, humanitarian action and development—A personal account.* Washington, DC: Conflict Prevention and Reconstruction Unit, World Bank.

World Bank. 2005. *Reshaping the future: Education and postconflict reconstruction.* Washington, DC: World Bank.

7   Sustainable Democracy
Issues, Challenges, and Proposals for Citizenship Education in an Age of Climate Change

Ian Davies and James Pitt

INTRODUCTION

Climate change is a political matter. This is true in the ways in which education for sustainability has been characterized, how and why climate change is occurring, and how we can think and act in ways that will help in the struggle for survival. This chapter will, in the context of climate change, refer to the nature of citizenship and discuss the meaning of citizenship education. It will explore the case study of England, where citizenship educators have according to analyses of official publications largely neglected education for sustainable development, suggest what should be done, and provide two snapshots of possible futures. Throughout we will highlight the key issues that need to be considered in order for the implications of climate change to be taken on board fully by citizenship educators.

WHAT IS CITIZENSHIP AND HOW DOES IT CONNECT WITH THE IMPLICATIONS OF CLIMATE CHANGE FOR SOCIAL LEARNING?

Of the very many issues that could be discussed about the nature of citizenship we will refer to only four. We will discuss where citizenship is located, what is expected of a citizen in terms of rights and/or duties, whether or not we can refer to a crisis in citizenship, and what sort of perspective on the environment is adopted by the citizen?

Citizenship and 'Location'

The term 'citizen' is often used very loosely but when dealing with climate change (a global concern), it is important to know what is being referred to. In many debates about citizenship it is not always immediately clear whether national citizenship or something else is being discussed. Heater (1997) has made several extremely valuable contributions to help us clarify these matters that are especially significant for anyone concerned about climate change. It is important to know whether a citizen seeking to influence climate change policies should be concerned with the operation of governments in, largely, national contexts or whether something else entirely is being suggested by the use of 'citizen.' Heater (1997, 22) outlines those forms of citizenship that exist beyond the nation state (Table 7.1).

Heater goes on to explain (1997, 36–38) that there is a range of meanings that can be applied to global citizenship. He outlines (37) four main meanings that can be placed on a spectrum of which the opposite ends are 'vague' and 'precise' (Table 7.2). Thus it is not enough for those concerned with climate change to call for citizen action. We need to know what sort of citizenship we are expected to exercise. This of course has major implications for the sort of education that we wish to advocate. Educating people to act in relation to a nation state is not *entirely* different from transnational matters but at Heater's vague end of the spectrum we will see something very different (and probably attitudinal focus) from those teachers who occupy a more 'precise' position (and would wish to emphasize understanding and action in relation to formal constitutional structures). These debates are of great significance with, for example, Held (1995, 279) promoting forms of global governance (e.g., reforming the UN Security Council to give developing countries a significant voice and decision-making capacity; enhanced political regionalization, including in the EU and beyond; and the use of transnational referenda) that are very relevant to climate change. Others such as Kiwan (2008) discuss forms of citizenship that rely less on political structures (e.g., forms of collaboration between and within women's groups).

Table 7.1   Forms of Citizenship Held in Conjunction with State Citizenship

Legally defined	Dual—citizenship of two states held simultaneously
	Layered—in federal constitutions; and in a few multinational communities
Mainly attitude: limited legal definition	Below state level —municipal, local allegiance/sense of identity
	Above state level—world citizenship

Table 7.2   The Meanings of World Citizenship

VAGUE			PRECISE
Members of the human race	Responsible for the condition of the planet	Individual subject to moral law	Promotion of world government

Citizenship: The Liberal and the Civic Republican Traditions

Our second issue about citizenship and climate change refers to expectations of—and for—the citizen. Broadly, two traditions exist within citizenship education: the civic republican, which principally relates to responsibilities exercised in public contexts; and the liberal, which, mainly, relates to the rights of individuals in private contexts. Liberal citizens might be expected to emphasize their right to make decisions about resources that they see as being garnered through their own enterprise. Civic republicans would be, generally, more open to the recognition of collective responsibility in public contexts. It would be naive and simplistic to see these traditions as being mutually exclusive. One can, of course, develop individual advantage from collective action; one can act individually in order to support the mass. Those who campaign for action in the face of climate change do not fit easily into the liberal or civic republican 'pigeon holes.' But these dimensions are useful for seeing the strands of thought contained within citizenship and this shows, again, that a call for 'citizens' to act in relation to climate change is inadequate. We need to reflect on the sort of citizen we are exhorting and the form of citizenship that we are promoting.

Citizenship and Crisis

Thirdly, we need to be aware of the context in which citizens think and act. It is entirely usual for the crisis card to be played. Citizenship education has been developed in recent years mainly because of a so-called crisis in democratic societies. To an extent we accept there is a crisis. Turnout is low in elections (the 2001 general election in England saw the lowest turnout since 1919 and the 2005 data were only slightly higher). In terms of community involvement or engagement we are said by some commentators to be 'bowling alone,' acting as individuals and contributing to a decline in associational activity of sports clubs, cultural groups, and political bodies (Putnam 2000). When considered in light of evidence of rampant climate change these claims of democratic crisis seem to suggest that irresponsible and apathetic citizens are acting (or neglecting to act) in ways that will damage the planet. However, it would be good to take a measured approach. We do not wish to suggest that the evidence for climate change is unconvincing. We are entirely persuaded that human action is responsible for significant environmental degradation and that urgent action is required. However, in the context of the nature and expression of citizenship we need to be cautious. Electoral turnout has always been low (Jeffreys 2007) and there is evidence of high levels of community engagement (Pattie et al. 2004). Young people are aware of environmental issues and often show that while they might be bored with politicians, they are interested in politics (Haste 2004). Our key consideration here is not to debate whether or not there is a crisis in society: there is, there always has been, and always will be. But in a changing situation we need to examine evidence carefully. For the purposes of developing an appropriate education in difficult circumstances we need to avoid the rush to judgment and to precipitate action that will not help. Crisis is not a good point from which to plan curricula, and unless it is avoided may well only lead us to demands for narrowly behaviorist training (Sears and Hyslop-Margison 2007).

Citizens and Perspectives on the Environment

Finally, and perhaps most fundamentally, we need to consider the perspectives that are brought to bear on and for the environment by the citizen. This requires consideration of the evolution of citizenship as well as its current condition. Marshall (1963) traced the history of citizenship by referring to the growth of civil rights (eighteenth century), political rights (nineteenth century), and welfare rights (twentieth century). More recently commentators have developed new perspectives that recognize both particular groups (e.g., feminist perspectives discussed by Arnot 2008) as well as the need for the rights of the environment to be secured (e.g., Curtin 2002), thus taking us beyond discussion of the rights of individuals and group rights and into new fields that see rights being held not only by people. There is also significant discussion about the perspectives of what some have called 'green citizens' (Campbell and Davies 1995). This requires consideration of attitudes towards environmental solutions. These solutions are cast either very generally (e.g., Huckle 1990; Vare and Scott 2007) to suggest either that science and technology can provide ways forward or that science and technology are themselves the causes of environmental problems. From these rather broadly based reflections on the nature of the connection between citizenship and the environment may be developed a range of actions. Currently three general (and not necessarily mutually exclusive) types of action by individuals seem to emerge from these perspectives: the acceptance of the need to recycle resources, an economic position in which consumer choice is seen as being influential, a political approach in which campaigning is seen as necessary to change policy and practice. In small-scale pieces of research (e.g., Campbell and Davies 1995) the last approach is mentioned and practiced least often by respondents but we feel that it is a significant key to a positive future. But rather than simply express our own preferences our most important point here is that we do need to be clear about the perspectives we expect of citizens and in light of this the sort of education we might develop. For this reason we move from a concern with citizenship to a more explicit consideration of citizenship education.

WHAT IS CITIZENSHIP EDUCATION AND HOW DOES IT RELATE TO CLIMATE CHANGE?

In one sense it is very easy to define or characterize the field of citizenship education. Heater (1990, 336) has explained that:

A citizen is a person furnished with *knowledge* of public affairs, instilled with *attitudes of civic virtue*, and equipped with *skills to participate* in the political arena [italics added].

It seems that Heater's views have been accepted by government. The current version of the National Curriculum in England declares:

Education for citizenship equips young people with the knowledge, skills and understanding to play an effective role in public life. (Qualifications and Curriculum Authority 2008a)

This is, obviously, a very important educational field. And yet one of the most noticeable features in the history of citizenship education has been its neglect. Although there were exceptions (e.g., see Heater 1984) little, if any, explicit citizenship education was considered necessary for all school-age students prior to about the mid-1970s (Entwistle 1973; Batho 1990). When practiced it occurred as 'academic' courses about constitutions for high-status students or civics for low-status students who were told the rules they had to follow. A mixture of positive and negative causes has led to an explosion of interest in citizenship education around the world. It is accepted that political messages are ubiquitous and capable of being understood by young people who, as the age of majority is lowered and the years of full-time education are extended, are seen by policy makers influenced by communitarian ideas as perhaps less than wholly enthusiastic, poorly informed voters and taxpayers. These pressures have seen initiatives for civic or citizenship education in many countries (see Arthur et al. 2008) with, for example, much work taking place across Europe (2005 was the European Year of Citizenship through Education) and elsewhere. This does not, of course, mean that citizenship education means the same in all these different locations. Davies and Issitt (2005) show through a study of textbooks that forms of citizenship education in England, Canada, and Australia are very different. It also does not mean that citizenship education is necessarily benign. Citizenship education may be aligned with dictatorship as well as democracy. Accusations of nationalism and the absence of open democratic debate in classrooms in China (Cheung and Pan 2006) or Singapore (Sim and Print 2005) or England (Leighton 2006) may easily be seen. These tensions are replicated and perhaps reinforced when one considers the origins and nature of environmental education with some, for example Branwell (1994), highlighting the unacceptable alignment of environmental purity with racism and political oppression in which "the 'biologic' point of view that saw man as one with nature had been part of the tradition encouraged by the Nazis" (43). What we need to do is not celebrate citizenship education as if it necessarily means good things but rather investigate closely where it is said to exist. In order to do this we need to examine particular contexts and in the next section of this chapter we provide a case study of one country, England.

CITIZENSHIP EDUCATION IN THE CONTEXT
OF CLIMATE CHANGE: A CASE STUDY

The National Curriculum for England is divided into several areas. Concepts, rightly in our view, have a place of central importance and three are particularly highlighted. They are:

- Democracy and justice
- Rights and responsibilities
- Identities and diversity: living together in the United Kingdom

But teachers and learners are encouraged, again rightly in our view, to consider the best ways to promote:

- Critical thinking and enquiry
- Advocacy and representation
- Taking informed and responsible action

We respect and agree with this approach. We do not wish to criticize the idea of a national curriculum that provides a guarantee of equal distribution of knowledge to all learners. A national curriculum is, broadly in our view, an acceptable means of democratizing the curriculum. The particular approach to framing the curriculum referred to earlier also seems to us a good way to develop learners' understanding of and involvement in contemporary society and avoids simply requiring students to remember information. Concepts and processes are more important than content. If we were to rely on content exclusively as a means of promoting understanding and developing the capacity for action we would be severely limited. Such an approach would lead immediately to problems as there would be no guidance as to what could be regarded as significant and we would be pushed on the ebb and flow of media panics and the short-term nature of government initiatives rather than tackling foundational matters.

However, the problems associated with a national curriculum have not been escaped entirely. Once the political step of establishing a national curriculum is taken the disadvantages associated with it become immediately apparent. There are, of course, issues about the perceived necessity of all national curriculum subjects becoming entrapped in what we want, critically, to refer to as academic paraphernalia. If one of the purposes, for example, of the National Curriculum is to judge the achievement of learners, then the examination system with associated league tables for schools and an interventionist and punitive inspection system takes shape.

However, we do not in the space of this short chapter intend to discuss all the issues concerning the formation and implementation of national curricula. Rather it is immediately obvious that the authors of the National Curriculum may not take the issue of climate change seriously. Several examples will be sufficient to demonstrate this point. Of the links that are proposed between citizenship education and education for sustainable development some fall into what could be regarded as a general category. In relation to key processes the National Curriculum documents assert that "pupils should be able to engage with and reflect upon different ideas, opinions, beliefs and values when exploring topical and controversial issues and problems." More specifically in the section that deals with "range and content" of citizenship education one of the eleven recommended study areas mentions the environment explicitly ("the study of citizenship should include actions that individuals, groups and organisations can take to influence decisions affecting communities and the environment"). In the section on "curriculum opportunities" one of ten recommended areas is for the "curriculum to provide opportunities for pupils to take into account legal, moral, economic, environmental, historical and social dimensions of different political problems and issues." In other words the environment is neither seen as a fundamental matter or as something that requires extensive illustration or elaboration. Citizenship education does not, in the version of the National Curriculum that applied to schools in England in 2008, especially concern itself with climate change.

It is worth noting that the National Curriculum for England *does* mention sustainable development in three other subjects—design and technology, geography, and science—and there are cross-curricular dimensions designed to "provide important unifying areas of learning that help young people make sense of the world and give education relevance and authority" (Qualifications and Curriculum Authority 2008b). These include identity and diversity, community participation, and global dimension and sustainable development—all areas relevant to citizenship education. It remains to be seen whether teachers address these cross-curricular dimensions actively and creatively, or whether they will reside only in subject bunkers. Furthermore the Department for Children, Schools and Families (DCSF) has a sustainable schools policy in which every school is invited (not required) to assess its behavior with regard to eight 'doorways' that include key environmental issues such as use of energy and water, purchasing and waste, buildings and grounds, food and drink, and travel and transport. Under the sustainable schools banner there are also doorways relating to inclusion and participation, local wellbeing, and the global dimension. Schools are asked to evaluate what they do in each of these doorways in the areas of curriculum, campus, and community (DCSF 2007). Although such injunctions for good practice have been in place for some time there is little evidence that they are widely adhered to (Ofsted 2008).

We entirely accept that formal policy documents are not always a good guide to what happens in practice. It is possible that teachers and learners could be, following young people's high levels of interest and engagement with the environment, developing lessons, schemes of work, and other activities that promote a critical and active engagement. However, if this is taking place we do wish to emphasize the scale of challenge that would need to have occurred in the face of the legitimized view of what sort of work should take place. Winter's (2007) analysis of policy documents is interesting:

> The policy documents examined conspire to prevent discussion of important political points, but at the same time they appear to address the environmental issues central to SD. They do this because of a rhetorical force: the documents provide—so the texts seems to say—exactly the kind of clarity and helpful, practical advice that teachers really need. But the reality is that this rhetoric obscures precisely those political and ethical questions with which any education for sustainability needs to engage. In consequence, the teachers receive no clear guidance for teaching about environmental issues, with the result that they either ignore ESD or deal with it in inappropriate ways, however well-intentioned these may be. Such well-intentioned teachers are likely themselves to be colluding in policies that confuse and prevent the central ethical and political issues coming to light. (350)

The limited success of education for sustainable development (ESD) in England can be attributed in part to the inherent contradictions between the sort of neoliberalism favored by the British government and social democracy (Huckle 2008). Such a dichotomy between rhetoric and reality is not new (Stevenson 1987, 2007). A number of authors (e.g., Bonnett 2003; Sterling 2005; Huckle 2006; Vare and Scott 2007) argue in favor of an approach to ESD in which argument and debate lie at its heart. If ESD is constructed as an arm of policy (simply informing learners about matters relating to sustainability) it will continue to mask the underlying tensions, powerful interests, and tensions that are leading humanity towards *un*sustainable development. We need to construct sustainability and hence education for sustainability as a 'frame of mind' (Bonnet 2003). This will not be easy—indeed it is a recipe for conflict within educational institutions. But it is more likely to engage young people as citizens.

WHAT SHOULD BE DONE?

What then needs to be done if we are to educate for citizenship in ways that recognize the challenges of climate change? We suggest that there is the need for the recognition and further development of political perspectives. We propose reliance on a broadly based conceptual approach that allows learners to explore environmental issues and develop knowledge, skills, and dispositions that are needed for a sustainable future. This demands an exploration of what "constitutes a right relationship with nature," which raises questions "not only about basic understandings of, and motives towards nature, but also about human identity and flourishing"; it will engage not just at a conceptual or cognitive level, but also at aesthetic, affective, imaginative, and moral levels (Bonnett 2003).

This will mean five priorities. First we must focus principally on contemporary matters. We do not underestimate the potential and power of history education to help learners understand key issues but we suggest that an explicit focus on the present and future will yield most benefits. There is some limited but very valuable work on futures education (e.g., Hicks 2002) and this needs to be developed further. Learners need to know and understand about issues that concern them now and need actively to prepare for—and to shape—what is to come.

Second, this material must, as we argued earlier, be based on more than content in the form of information. Rather there should be a conceptual core to the curriculum. These concepts should be of at least two types: the substantive and the procedural. Substantive concepts are those that most readily suggest content. A concern with power, for example, will lead teachers to focus on expressions of power in contemporary society by politicians or others. Procedural concepts are also vitally important but are somewhat harder to clarify. Procedural concepts frame how study takes place. For example, learners have to understand the nature of evidence as they engage with citizenship education. To 'do' citizenship one must handle evidence. If we know what conceptual processes underlie learners' work then we may be more easily assured that significant matters are being addressed. We do not wish to claim at this stage that we know precisely all the concepts that should be used to frame a curriculum appropriate for social learning at a time of rampant climate change but we do wish to argue that there is a need for work in this area that will aid identification of these matters.

Third, the learning that takes place must occur in an appropriate context. It seems unlikely that lecturing young people about what they must do in order to save the planet will have more than a temporary shock value. If we want learners to understand how to take action then a considerable part of their education must occur in an open climate in which engagement is genuinely valued. Our argument to avoid a situation in which teachers expect thought without action does not mean that we embrace the simplistic polar opposite: thoughtless action. The key priority for educators, of course, is to educate. Schools are political institutions and we want students to see themselves as people who act, but we risk wasting young people's time if we preach to them about the actions that they should take. We need to encourage action and, just as important for the learning process, reflection on action.

Fourth, as educators we have a responsibility to consider the impact of our professional work. We need to explore what young people know and can do and refine their and our teaching and learning initiatives. This call for assessment does not mean that we are arguing for a narrow testing regime. It does mean that we should accept the inevitability of students and teachers making judgments about each others' work and suggesting that there is a need for this to be done explicitly in a way that helps the creation of the educational work that we desire (Jerome 2008).

Finally, we suggest that citizenship education involves the pursuit of justice. It is not a neutral enterprise. While respecting and requiring freedom of thought and speech and action educators are concerned to do the right thing. There should not be, in Crick's memorable phrase, the "postmodernism of the streets" (Crick 2000, 125) in which anything goes. Some things are better than others. Some ideas are better than others. It is better that the planet survives than dies. Democracy is better than dictatorship. The task for the educator is to investigate what those things mean and to help learners to achieve positive ways forward.

WHAT WILL HAPPEN IN THE FUTURE?

Lovelock (2009) predicts that it is already too late to prevent catastrophic climate change. He criticizes the smooth paths of slow, sedate change that are postulated by the Intergovernmental Panel on Climate Change, suggesting rather that we will be confronted by sudden lurches as positive feedback mechanisms kick in—whether they be in loss of albedo at the poles, acidification of the oceans, or abrupt changes in the circulation of cold and warm water in the North Atlantic. Lovelock argues that citizens and governments should stop looking to moral palliatives such as buying low energy light bulbs and installing wind turbines, or eating locally produced, organic food—all behaviors that schools are being urged to adopt. He ridicules planting trees as a way of offsetting the CO_2 emissions of air travel, comparing this to the indulgences sold by the Roman Catholic Church to wealthy people who wished to reduce the time that they would inevitably spend in purgatory.

Lovelock argues instead that governments need to consider how best to mitigate the social, economic, and political chaos that is bound to come. He predicts that there will be a number of "lifeboat islands" in which the rump of the human species will struggle to survive, and suggests that the ethics of living in a lifeboat will be totally different from the "cosy self indulgence" of the latter part of the twentieth century.

What might the ethical issues be, and how might these affect what is currently called citizenship education?

Let us speculate within Lovelock's grim scenario in which he predicts that the British Isles will become such a lifeboat island—or rather a group of islands as rising sea levels annihilate most of the coastal areas and the existing population retreats to the hills. There is not enough land to accommodate everyone, let alone the climate change migrants who have fled the deserts of Spain, Italy, and southern France. Following fierce debates about leaving the European Union with its free movements of citizens before 'they' swamp 'us' completely it rapidly becomes necessary to redefine legal citizenship. These debates start in the first decade of the twenty-first century as the world recession generates unemployment and politicians champion the cause of 'British jobs for British workers.' Citizenship ethics polarize into accepting either that the prime role of the state is to protect its own, or arguing in favor of a common humanity. Other ethical discussions shift focus into such areas as whether it can ever be morally acceptable to depart from the compulsory euthanasia, sterilization, and abortion that now prevails, all policies demanded by the public good as food and water becomes scarce.

Small anarchist or religious groups try to maintain an ecomonastic lifestyle based on the values of the sanctity of human life and the right to self-determination. But pressure in the lifeboat increases and these groups become increasingly marginal; crude survival utilitarianism predominates. Schools and universities as they were known in the early twenty-first century have all but disappeared; the former are called 'community learning centers' and are devoted to producing model citizens, the latter are research centers seeking rapid advances in geo-engineering, food production, and desalination. Indeed most scientific endeavor is devoted to this cause.

Most learning is driven by all-pervasive digital technologies, and face-to-face teacher-student interaction has largely disappeared. What were called 'information and communications technologies' at the start of the century are now just called 'coms.' These give people instant access to all the data on the Web, using voice-activated software to start with, and then microchips installed in the brain that allow the user to call up data using carefully channelled thought searches. However, much of the vast amount of data is 'planted' by governments who seek to shape the way that young people think (this was brought in under the banner of 'personalized learning' in the first decade of the century). Indeed governments seek total control of access to the Web but anarchist programmers stay one step ahead and for those who wish to find alternative viewpoints, the information is there.

During this shift, concepts of 'citizenship' and 'belonging,' 'rights,' and 'responsibilities' increasingly dominate educational debate. Some authorities in education argue that formal institutions must play an ever-increasing role in socializing young people into seeing and understanding their place in a changing world. This role is determined by planners whose main responsibility is to maintain the social and economic health of the nation. Other educators take to the streets so to speak; they decide that education still has a role in developing learners as creative, autonomous, critical beings, but see that this is impossible within the formal education sector. They work with pressure groups and the flourishing networks of resistance cells. For them 'research' is devoted to developing small and cheap technologies that make it possible to avoid 24/7 surveillance, and to grow food.

As long as there is still hope that Lovelock will be proved wrong and that planetary disaster can be avoided, there is an emergence of 'transition educators.' These people build on the foundation of the transition towns such as Totnes that emerged at the beginning of the century (see Transition Town Totnes 2009). This is a citizen-based movement in which people attempt to develop their own communities in a sustainable way. The new transition educators begin to ask how educational institutions might develop if we are to take seriously the triple shocks of climate change, peak oil, and profound global economic uncertainty.

To depart temporarily from the delights of unbridled speculation, it is worth referring to Webster and Johnson (2009), who provide an unusual and prescient way forward for the early stages of transition education. They raise questions as to what might educational institutions look like if they mimicked nature, in which there is no waste. They call for a paradigm shift in the way that we think, moving from a linear or throughput mentality with its attitude of 'take–make–dispose' towards systems thinking and cyclical thinking with the need to 'close the loop.' In this the 'waste' of any process is not wasted but becomes the feedstock of some other process. They point to good practice in industry where this model is the basis of development, such as InterfaceFlor. This is a carpet service which will collect your worn carpet and recycle it, while recarpeting your building with 'new' carpet made from other recycled carpets. They quote Ray Anderson, the founder and chairman of Interface Inc., describing him as the visionary behind the change in culture of the business:

> We can look to nature for the inspiration, the guiding principles to make the changes needed in the industrial system to make it as effective as nature is—waste free, resource effective and resource efficient, benign, operating on sunlight the way nature operates on sunlight; taking nothing and doing no harm. (79)

Other examples in Webster and Johnson's call to rethink education are the ecocities in Masdar, Abu Dhabi, and Dongtan, China, with their positive energy buildings that generate more energy each day than they consume. The authors ask if it is not possible that educational planners can learn from industry as well as nature? The final three chapters of this provocative book attempt to answer this by looking at learning and transition, the sustainable school or college of the future, and why worldviews matter. They conclude that the bottom line is about citizenship:

> [Sustainability] is a citizenship question first, and not a consumer one. The balance of work in schools and colleges needs reviewing so that it is *more* about 'systems and citizenship' and *less* about 'me and consumerism.' (145)

For Webster and Johnson the development of citizenship education and education for sustainable development are inseparable.

We would like to return to the two speculative scenarios for the immediate future. In one scenario (closer to Lovelock's dire warnings) we continue to develop as an individualistic, liberal, science-based society, in which citizens do not need to think of themselves as part of a democratic collective. The rights of an individual take precedence over the need for a coherent approach in which there is recognition of the 'rights' of the environment. The commonsense view prevails that what one can avoid paying for is not seen as being paid for by anyone. Both citizenship education and education for sustainable development are prescribed by curriculum authorities with a well-planned content in the hope that knowledge about the carbon cycle, greenhouse gases, environmental degradation, etc. will somehow induce radical behavior change at individual and collective levels. It is education *about* sustainable development and the rights and duties of citizens. But there is little or no encouragement of more radical dialog and debate, or for learners and their teachers to explore the contradictions in the way their educational institution is managed. The issues remain objects to be studied. Education does not become critical. In this context environmental degradation continues and accelerates, albeit not in a linear fashion. There is much championing of products and technologies that allow the status quo to remain unchallenged—students learn about 'green cars' (an oxymoron if ever there was one) and carbon-trading or carbon-offset schemes. They recycle their rubbish and buy more fair-traded or organic products. Individualist assumptions about self-actualization and the nature of fulfilment remain intact.

Alternatively we might develop a civic republican approach that is meaningful. Peterson (2008) recognizes the distinctions between instrumentalism and intrinsic republicanism. In this scenario the rights of the individual and the needs of groups are recognized. Both citizenship education and ESD build on a coherent approach to group action that benefits individuals. ESD moves beyond education *about* sustainable development and possibly beyond education *for* sustainable development, and is located within education *through* sustainable development (Pitt and Lubben 2009). In this way the division between school and community disappears as learners are encouraged to engage in social action and to question everything. Indeed both citizenship education and ESD as subjects or prescribed curriculum areas disappear as learners, teachers, parents, and stakeholders throughout the community take more control of curriculum and pedagogy. Educational institutions become both a focus and locus for action and reflection on sustainable change. This is not a call for anarchy; control is for a purpose and is commensurate with desired outcomes. In this scenario there is a humanization of education, a recognition that content, process, and outcomes are in a critical, dialogical, and developing relationship with a developing understanding of humanity and its right relationship with nature. Accompanied by changes in lifestyle, a cyclical or closed-loop paradigm of development, and whatever geoengineering that is needed, this leads to both a healthier citizenry and a healthier planet. Part of this health is manifested by a widespread 'strategic optimism' (Webster and Johnson 2009, 146). But educational institutions and processes as we know them will have been replaced by something very different.

REFERENCES

Arnot, M. 2008. *Educating the gendered citizen*. London: Routledge.

Arthur, J., I. Davies, and C. Hahn, eds. 2008. *Sage international handbook of democracy and citizenship education*. London: Sage.

Batho, G. 1990. The history of the teaching of civics and citizenship in English schools. *The Curriculum Journal* 1 (1): 91–100.

Bonnett, M. 2003. Education for sustainable development: A coherent philosophy environmental education? *Journal of Philosophy of Education (Special issue: Retrieving nature: Education for a post humanist age)* 37 (4): 675–90.

Branwell, A. 1994. *The fading of the greens: The decline of environmental politics in the west*. New Haven and London: Yale University Press.

Campbell, R. M. and I. Davies. 1995. *Education and green citizenship: An exploratory study with student teachers*. York: University of York.

Cheung, K. W. and S. Pan. 2006. Transition of moral education in China: Towards regulated individualism. *Citizenship Teaching and Learning* 2 (2): 37–50.

Crick, B. 2000. *Essays on citizenship*. London: Continuum.

Curtin, D. 2002. Ecological citizenship. In *Handbook of citizenship studies*, ed. E. F. Isin and B. S. Turner, 293–394. London: Sage.

Davies, I. and J. Issitt. 2005. Reflections on citizenship education in Australia, Canada and England. *Comparative Education* 41 (4): 389–410.

DCSF. 2007. *Sustainable schools—National framework*. http://www.teachernet.gov.uk/sustainableschools/framework/framework_detail.cfm (accessed December 11, 2008).

Entwistle, H. 1973. *Political education in a democracy*. London: Kegan Paul.

Haste, H. 2004. *My voice, my vote, my community: A study of young people's civic action and inaction*. London: Nestlé Social Research Programme.

Heater, D. 1984. *Education for peace*. Lewes, DE: Falmer Press.

———. 1990. *Citizenship: The civic ideal in world history, politics and education*. London: Longman.

———. 1997. The reality of multiple citizenship. In *Developing European citizens*, ed. I. Davies and A. Sobisch, 21–48. Sheffield: Sheffield Hallam University Press.

Held, D. 1995. *Democracy and the global order: From the modern state to cosmopolitan governance*. Cambridge: Polity Press.

Hicks, D. 2002. *Lessons for the future: The missing dimension in education*. London: Routledge.

Huckle, J. 1990. Environmental education: Teaching for a sustainable future. In *The new social curriculum: A guide to cross-curricular issues*, ed. B. Dufour, 150–166. Cambridge: Cambridge University Press.

———. 2006. *Education for sustainable development: A briefing paper for the Teacher Training Resource Bank (TDA)* (updated edition). http://john.huckle.org.uk/publications_downloads.jsp (accessed November 20, 2008).

————. 2008. An analysis of New Labour's policy on education for sustainable development with particular reference to socially critical approaches. *Environmental Educational Research* 14 (1): 65–75.

Jeffreys, K. 2007. *Politics and the people: A history of British democracy since 1918.* London: Atlantic Books.

Jerome, L. 2008. Assessing citizenship education. In *The Sage international handbook of democracy and citizenship education,* ed. J. Arthur, I. Davies, and C. Hahn, 545–58. London: Sage.

Kiwan, D. 2008. Friends, Londoners and countrywomen: From women's practices of 'sociality' to inclusive participative citizenship? Paper presented at Britishness, Identity and Citizenship: The View from Abroad conference, June 2008, at University of Huddersfield.

Leighton, R. 2006. Revisiting Postman and Weingartner's 'new education'—Is citizenship education a subversive activity? *Citizenship Teaching and Learning* 2 (1): 79–89.

Lovelock, J. 2009. *The vanishing face of Gaia.* London: Allen Lane.

Marshall, T. H. 1963. *Citizenship and social class.* London: Heinemann.

Ofsted (Office for Standards in Education, Children's Services and Skills). 2008. *Schools and sustainability: A climate for change?* London: Ofsted.

Pattie, C., P. Seyd, and P. Whiteley. 2004. *Citizenship in Britain: Values, participation and democracy.* Cambridge: Cambridge University Press.

Peterson, A. 2008. *The civic republican tradition and citizenship education.* [PhD dissertation, University of Kent.]

Pitt, J. and F. Lubben. 2009. The social agenda of education for sustainable development within design and technology: The case of the Sustainable Design Award. *International Journal of Technology and Design Education,* 19(2): 167–86.

Putnam, R. D. 2000. *Bowling alone: The collapse and revival of American community.* New York: Simon & Schuster.

Qualifications and Curriculum Authority. 2008a. *The National Curriculum Citizenship Key Stage 3.* http://curriculum.qca.org.uk/key-stages-3-and-4/subjects/citizenship/keystage3/index.aspx?return=/key-stages-3-and-4/subjects/index.aspx (accessed February 12, 2009).

————. 2008b. *The National Curriculum—Cross Curriculum Dimensions.* http://curriculum.qca.org.uk/key-stages-3-and-4/cross-curriculum-dimensions/index.aspx (accessed December 11, 2008).

Sears, A. and E. Hyslop-Margison. 2007. Crisis as a vehicle for educational reform: The case of citizenship education. *Journal of Educational Thought* 41 (1): 43–62.

Sim, B. Y. and M. Print. 2005. Citizenship education in Singapore: A national agenda. *Citizenship Teaching and Learning* 1 (1): 58–73.

Sterling, S. 2005. *Whole systems thinking as a basis for paradigm change in education: Explorations in the context of sustainability.* [PhD dissertation, University of Bath.] http://www.bath.ac.uk/cree/sterling.htm (accessed July 6, 2009).

Stevenson, R. 1987. Schooling and environmental education: Contradictions in purpose and practice. *Environmental Education Research* 13 (2): 139–53. Orig. pub. 1987.

Stevenson, R. 2007. Schooling and environmental/sustainability education: From discourses of policy and practice to discourses of professional learning. *Environmental Education Research* 13 (2): 265–85.

Transition Town Totnes. 2009. http://totnes.transitionnetwork.org/ (accessed February 29, 2009).

Vare, P. and W. Scott. 2007. Learning for a change: Exploring the relationship between education and sustainable development. *Journal of Education for Sustainable Development* 1 (2): 191–98.

Webster, K. and C. Johnson. 2009. *Sense and sustainability—Educating for a low carbon world.* Preston, UK: TerraPreta.

Winter, C. 2007. Education for sustainable development and the secondary curriculum in English schools: Rhetoric or reality? *Cambridge Journal of Education* 37 (3): 337–54.

8 School Improvement in Transition
An Emerging Agenda for Interesting Times

Jane Reed

By every . . . lesson I was taught to think of myself as somehow apart
from soil and dust. Very little in my culture taught me that my existence depends on the existence of earth.

(Griffin 1996, 75)

Our species is at a crossroads. Time is running out. We must re-invent
the way we are living on earth. Education and learning are a deep and
essential part of this change.

(Fox 2006, 10)

INTRODUCTION

The ecological, social, and economic emergency of our civilization takes
form now as climate change. Climate change is the main theme of this
book although as pressing for attention are related forms of planetary
emergency that include poverty, disease, crime, and violence. Despite the
wake up calls of the past thirty years there is only a tenuous link made
in the collective mind maps of the West between our dependency on the
biosphere and the separation from it that results in the way we live and
think. This contradiction is more evident now than at any time since the
start of the industrializing process and in the gap, the ebb and flow of this
paradoxical situation, is the existence of the human species and all that
we have become. The possibility to learn from our mistakes and evolve
hangs in the balance. Climate change is one of the resulting threats and
challenges of our time.

Lester Brown suggests that if we value human life the changes that will
secure its future need managing with unprecedented speed (Brown 2008).
Writers and commentators (Bowers 2001; Orr 1993; O'Sullivan 2001; Sterling 2001) suggest education and in particular, learning, will either have a
leading role in taking us out of the crisis or they are going to prevent us
from doing so.

This chapter asks whether the state-funded education system in England
has either capacity or grip on the impending climate crisis to provide its students with the knowledge, strategies, and motivation to meet the challenges

of this period of human history. It asks what it will take to mobilize the theory and practice of the educational vision and response that is required. At the moment there is as broad a range of understandings of what a greater environmental emphasis for the school system might involve as there are scenarios portrayed for climate change. In particular the readiness of the field of school improvement to provide ways of framing the thinking and rapid adjustment and changes needed is questioned:

> We face unprecedented change . . . if suitable sustainable reform is to be achieved then education will play a major part in the transformation and in that process it too will be transformed and education will emerge as a stronger, more value based expression of lifelong processes of learning, activity and reflection which will better serve all those involved. (Clarke 2000, x)

Young people have a right to learn about climate change and sustainability, and given the opportunity they are motivated to do so. This is one of the reasons why the United Nations declared 2005–14 as a Decade of Education for Sustainable Development, which provides a contextual framework for learning about climate change although the use of the terminology in the field is deeply contested (Selby 2006). Young people were interviewed in a recent survey (Forum for the Future/UCAS 2007), which outlines their concerns about the world they are being brought up in. They identified a need for their schooling to provide them with ways of engaging with their world and its problems in a more authentic way than the rather fragmented, abstract encounter they have with her in traditional schemes of work and textbooks.

Since school attendance became compulsory almost a century and a half ago much has been invested in the form of 'school' and it is likely to continue to be the primary educational institution for most young people; particularly in urban areas as long as there is adequate longer-term energy descent planning. More diverse models of schooling will probably emerge as the Transition Town[1] movement gathers speed in the United Kingdom (Hopkins 2008) and as climate change speeds up. There is sadly neither time nor resource to redesign all our public educational institutions in ways that lend them more readily to the task.

The chapter reflects on current UK national developments. It suggests that what is needed is a more explicit focus on ecocultural sustainability for state schools that equips children and young people with the capacity to learn as well as the knowledge, resilience, and know-how for a world that faces the destabilization of climate change. The relentless, instrumental drive for educational standards could be put to better use if it was connected to a broader view of what it means to educate for wellbeing, democracy, citizenship, and sustainability rather than the current narrow goal of basic literacy and numeracy. A change of direction in school development efforts could enable public education to be emancipatory and take a lead in dealing with the anticipated turbulence of climate change.

Cuban (1995) identifies five different reform clocks—media, policy-maker, bureaucratic, practitioner, and student learning time—ticking away inside school reform that operate in descending order of speed. Ticking faster than them all is ecological time, a frame of time if taken seriously by educators would quickly alter the focus of the school system.

The context for this chapter is the Sustainable Schools Programme that is developing in England. This is a program to promote sustainability on the ground in schools by the government. Associated to it is a research and development program set up by the National College for School Leadership (NCSL) in 2007, which currently has over fifty schools involved in leading a local community of schools in promoting practices to develop their students' involvement in greater environmental and social sustainability and studying the leadership processes involved. The program challenges traditional models of school improvement that treat education for sustainability as a sideline while debates about the way forward remain predominantly in the academic domain. Some of the main findings of the research study to support Sustainable Schools (NCSL 2007) and examples from the NCSL Sustainable Schools Grant Scheme project will illustrate a transitional agenda for school improvement. The chapter argues that while there is still time social learning and school improvement theory and practice need urgently to learn from each other to provide a way forward for school reform.

THE CURRENT STATE OF THE FIELD
AND ITS CONTRIBUTION

School improvement as a field of enquiry began life in a group of European universities in partnership with the Organisation for Economic Co-operation and Development (OECD) in the late 1970s. The primary impulse in the field was to understand the organizational, social, and psychological processes needed to implement educational change and how these can clarify goals, manage complexity, and promote innovation. An early definition of school improvement came from one of the original OECD projects in the field:

> A systematic, sustained effort at change in learning conditions and other related internal conditions in one or more schools with the ultimate aim of accomplishing educational goals more effectively. (Van Velsen et al. 1985, 48)

The founding fathers of school improvement were preoccupied with the ubiquity and pace of change in the West in the postwar period. Young people, it was argued (ibid.) needed skills and knowledge to participate and work in an increasingly complex and globally connected world; school systems found themselves in a changing and uncertain context and required new support strategies. The field of school improvement began to examine how a school as an organization could become proactive, responsible, and autonomous (Hopkins 1987) as a way of dealing with change. The perceived shift was from a school as a collection of isolated, individual teachers to the enhancement of collective capacity to diagnose and solve problems. The founding energy was alert to context, optimistic, had a belief in the efficacy of a school as an organization, and envisioned new practices. It was a transformative, empowering agenda in the making. An agenda that foundered however in its original form (alongside other radical educational programs of the 1970s and 80s) as the political initiatives and policy-driven agenda of the post-1988 Education Act, heralding more centralized government control of education, commandeered school improvement for its own purposes and tamed its radical tendencies with technical responses that were never originally intended. Playing to the tune of political whim and fancy for two decades has left the theory and practice of school improvement weakened in its capacity to respond to something as life threatening as climate change.

Consequently school improvement as a field of both study and practice has not had the opportunity to evolve and has become stuck in a discourse that stopped short of critiquing its own materialist worldviews (Lodge and Reed 2003). As the school became increasingly the unit of focus, environmental degradation and climate change did not become priorities in the busy life of the system. The science of school improvement is a social one and has little concern with ecology: it lacks the sharp focus on context, sense of relation to place, local need, and learning from and with other fields of enquiry that is emphasized in education for sustainability.

In order to meet the pressures of policy requirements and the inspection of their implementation the improvement agenda has gradually became synonymous with the best student assessment outcomes, and the field has largely lapsed into 'business as usual' modes of operation that are rather heedless of the academic critiques of what has happened to schooling over the past twenty years. Resilience in the system is publicly defined by test scores.

Wrigley (2003) has observed that there have still been advances in the past two decades that build on the early work and founding energy in the field and that can contribute to the social learning needed to face climate change. These include developing collegial school cultures, the resulting theory and practice of 'distributed' leadership, supporting the learning of staff, the processes of school self-evaluation and the advances being made in student leadership and participation (Lodge 2008). Also the knowledge about how schools develop as learning systems is much more advanced now than even a decade ago (Sackney and Walker 2006).

School improvement as a field has become tired and lost its power partly because its players tend to talk to each other rather than across a wider constituency. David Orr (2008) has argued in his work that *laterality* will be a key process in tackling climate change. One of the things he means by this is academics, policy makers, and service providers speaking to each other across their current working silos. Designers need to talk to those working on renewable energy, food growers need to be in dialog with urban planners, water engineers need to talk to school architects. In policy terms the section of the government department concerned with the Sustainable Schools agenda may or may not be talking to the section of the same department concerned with school standards and effectiveness in a way that will promote more widespread laterality and coherence on the ground. Clarke (2009) notes that allowing diverse views can be difficult and that at times of transition new thinking is unwelcome. Incoherence continues as past meaning is imposed on present situations.

So although there are some innovative practices in place and advances in the theory of how to improve statewide education in England, sustainability is viewed as a topic not a central focus, is still seen predominantly as a cottage industry for the converted, and the school reform and improvement agenda in government is seemingly ambivalent about whether or not to give the system a policy mandate to promote sustainability. Despite the sustainable action plan of the Department for Children, Schools and Families (DCSF) (DFES 2005) the main educational policy arm of the government, and the activity of a few schools, the Office for Standards in Education (OFSTED), the inspecting arm of government, has recently reported (2008) that only a percentage of schools have sustainability on their agenda in more than a partial way:

> Most of the schools visited had limited knowledge of sustainability or of related initiatives. Work on sustainability tended to be piecemeal and uncoordinated, often confined to extra-curricular activities and special events rather than being an integral part of the curriculum. Therefore its impact tended to be short lived and limited to small groups of pupils. (4)

There is a view that the DCSF is reluctant to regulate for further large-scale change, fearing that compulsion will have a negative impact. The system is known to be creaking already under the effect of continuous new initiatives. Recent publications by both the Specialist Schools and Academies Trust (SSAT 2007) and the National College for School Leadership (NCSL 2007) have attempted to profile sustainability, if not climate change directly, but the scale of impact is still small and only implicitly connected to the implications for school improvement.

WHAT THE FIELD WOULD NEED TO DO
TO RESPOND MORE ADEQUATELY

Lester Brown in his *Plan B 3.0* (2008) outlines the task of saving civilization as stabilizing population, restoring the earth, feeding eight billion well, designing cities for people, raising energy efficiency, and turning to renewable energy. Education in an age of climate change goes beyond a narrow environmental education agenda. It requires a larger sense of purpose, one where community transformation and environmental concern are connected to each other and to the purposes of education. The agenda includes people, their habitat and health, as well as their connection to the biosphere (Reed 2008). The Transition Town movement (Hopkins 2008) leads the way in suggesting that tackling climate change is going to need locally based solutions including energy and food production and new transport and waste disposal systems.

In his work on social innovation Mulgan (2007) suggests that we need to understand more about how ideas move across and between sectoral boundaries. Developing David Orr's view of the importance of laterality, a first step on the path for greater social learning is for the fields of school improvement and education for sustainability to be in dialog, shared development activity, and to debate some of the lack of conceptual clarity (Selby 2006). Neither field is equipped on its own to provide the intellectual and pedagogical capital needed to build and mainstream educational models and practices to help mitigate climate change. This partnership could give a new perspective for the school system and provide a 'Plan B for Education' including the energy descent planning that will keep schools open and viable as a community resource.

A lateral partnership between sustainability education and school improvement could bring social learning and school improvement, two theories with much to say to each other, into better relationship. Glasser (2007) suggests that social learning results from:

> non-coercive relationships that rest on building a common language, transparency, tolerance, mutual trust, collaboration, shared interests and concern for the common good. (52)

Citing Glasser in the same volume Wals and van der Leij (2007) add to this understanding, describing it as:

> an interactive participatory, negotiated approach to, or process for guiding collective problem solving and decision making, that incorporates innovation, diffusion, systems theory and systems learning. (20)

These clarifications of the territory of social learning come close to what the spirit of the first school improvers had in mind for the primary discourse for school improvement.

In the research review undertaken for NCSL in their Sustainable Schools research (NCSL 2007) four main, interconnected improvement strands were identified in the school system that could provide the basis for the 'Plan B for Education.' This is also an exemplification of school improvement thinking and social learning theory coming together. The changes require not only the beliefs and priorities commonly identified in sustainability thinking but the organizational design and operating procedures that are identified by school improvement programs. They also provide examples that illustrate the OFSTED (2008) findings about what is needed (see also Figure 8.1):

1. Sustainability for climate change is still more in the *personal* than the *collective* realm. It is the initiative and preference of committed individuals or head teachers rather than the development of a whole system approach to the education of our young people.
2. At the moment it is a *partial* initiative—taking place in some classrooms and some schools, it isn't yet *integrated* into the whole system in a way that it will need to be if education is to confront, teach and tackle issues related to climate change.
3. This results in sustainability being an *additional* part of what a school has to do rather than *fundamental* to its goals and operating procedures
4. As a result of 1–3, schools, like the majority of public services, have an *inward* rather than an *outward* orientation to their core business.

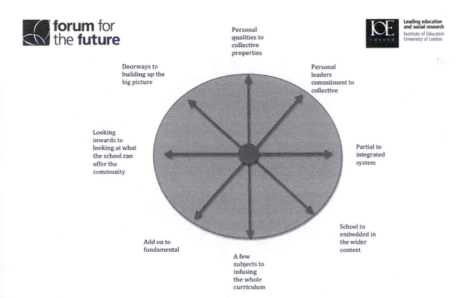

Figure 8.1 Eight processes for achieving a whole-school approach to sustainability (NCSL 2007).[2]

To enable these four school improvement strands for sustainability to take a firmer hold, the NCSL research study suggests that school improvement and leadership development would involve three main activities:

- Developing a whole-school approach that goes beyond the current "Eight Doorways" of Sustainability identified in the government's policy agenda for schools (DFES 2005) in which sustainability is everyone's role (Reed 2008)

- Redesigning the school curriculum so that sustainability, its goals, content, and processes are running more strongly through all subjects, the study of which extends beyond the school gates

- Enabling schools to become partners in a wider, less-isolated community, public service, and bioregional network[3] where adults and young people are learning and generating knowledge about sustainability together

For these changes to take place this chapter now identifies activity that will underpin this transition, aiming to give back to school improvement some of its transformative power by engaging it in approaches to social learning and preparing young people for the various scenarios of climate change. Two main aspects are identified: the development of an ecological worldview and the strengthening of learning orientation.

Developing an Ecological Worldview

Glasser (2007) strengthens the conceptual lack of clarity in the field about the nature and goals of sustainability identified by Selby (2006). He refers to 'ecocultural sustainability' (36) as the processes by which the human community can renew itself with governance, economy, and production that is sufficient, equitable, and learns and works with and from nature. This is a definition that could underpin public sector action to mitigate against climate change and avoid the confusion with 'development' that is in common usage. Ecocultural sustainability requires education and learning that recognize the place of the human in the biotic community, our reciprocal relationship and interdependence. It requires the shift from an industrial to an ecological worldview in school aims, organization, and pedagogy.

School improvement has a track record for identifying the aims for a school, deciding how to achieve them, and then evaluating progress towards realizing them. It has also developed approaches to learning and leadership for the whole stakeholder community. It is weak on conceptualizing and critiquing worldviews underpinning its activities (Lodge and Reed 2003) and is in default mode to an unexamined view of what a knowledge economy is without noticing how unsustainable such a worldview is. School improvement processes need to reflect the goals of sustainability as well as new theories of organization emerging from the field of ecological thinking (which foreground systemic interdependence and feedback from the environment), ecocultural design, and social innovation theory. If it is to have the capacity to contribute, school improvement rapidly needs to take ecological thought and values centrally into its discourse, develop an ecological worldview, a view of sustainability that is more than just maintainability, and infuse the system with a greater learning orientation. It will also need to build resilience and the capacity to contract, not just develop (Selby 2007).

Hargreaves and Fink (2005) are among a few writers in the school improvement field who draw on ecological metaphors. They comment on the need to *"take our heads out of the sand, the supermarket or the SUV"* (ibid. 3) and to note our part in the patterns of consumption, production, and standardization that have affected education. They suggest that:

> The environmental movement and its commitment to sustainability teach vital lessons for achieving sustainability in education organisations, the value of rich diversity over soulless standardisation, the necessity to take the long view, the wisdom of conserving resources, the moral obligation to consider the effects of our improvement efforts on others in the environment around us, the importance of acting urgently for change while waiting patiently for results, and the proof that each of us can be an activist and that all of us can make a difference. (ibid. 4)

The work of Sackney and Mitchell (2000) and Fullan (2005) alert the field to new theory for school improvement and the importance of thinking holistically rather than mechanistically. These writers however remain in the anthropocentric paradigm and stop short of connecting their ideas to an ecological worldview, a more explicit agenda about ecocultural sustainability, or the need to work with educators from other disciplines.

Beare (2006) takes thinking in the field further. His work suggests more explicitly that the "social imaginary" (worldview) needs to change because schools have reflected the logic and linearity of the industrial process in a way that is unecological: trapped as it is in standardization, specialization, synchronization, and centralization. School improvement can be criticized for insufficiently challenging these primary assumptions or questioning alternative paradigms. Beare draws on living systems theory, develops network theory as the basis for a new 'imaginary' for schooling, and critiques the image of a school walled off from its community and not promoting biodiversity.

> The question for educators, parents and the present generation of school children is whether their schools and schooling processes are

perpetuating an obsolescent imaginary which in many respects is be-
queathing consequences which are lethal, literally as well as figura-
tively. (52)

The NCSL research project (NCSL 2007) suggests that the value-base
needed to underpin activity in sustainable schools and to give them coherence is one that reflects an ecocultural approach to sustainability and
includes:

- A global sense of place

- Humans as integral to and inextricably interconnected to their context
  (i.e., our relationship to both the social sphere and the biosphere)

- The needs and rights of future generations

- Environmental justice

Theory to support a change of worldview and that supports the organizational change process for schools is provided by most of the key writers
in the field of environmental transformation and change and is increasingly used as the basis of change efforts for sustainability. It needs to be
more explicitly drawn on by the field of school improvement. Much of it
is based in the field of cybernetics that has become known in the environmental literature as 'systems thinking' and is a central strand in social
learning theory. Systems thinking emphasizes relationships and interaction rather than linear causality and means/end thinking. Sterling (2001)
and Tilbury (2008) see systems thinking as a key component of learning-
based change for sustainability. Systemic thinking is a particular feature
of the educational work of Senge (2006), who is one of the few writers in
the field of organizational change who has written and talked extensively
about the way in which school learning needs to change for sustainability.
He states:

> Human wisdom expressed in many deep-rooted and varied societal
> traditions around the world has long acknowledged that an under-
> standing of systems in their totality is the only foundation for making
> sound choices that benefit the health of the whole. (ibid. 18)

Senge (2000) notes the obsolescence of the system, that the patterns of
social organization that pervade many schools are still nearer to an industrial than an organic or ecological model and that schools have evolved
in relation to the needs of an increasingly industrialized and mechanized
society rather than their environment:

The result of this machine age thinking was a model of school apart from daily life, governed in an authoritarian manner, oriented above all to produce a standard product, the labour input needed for the rapidly growing industrial-age work place—and as dependent on maintaining control as the armies of Frederick the Great. (31)

He continues:

Other countries had their own local, indigenous texts, both written and oral. They learnt about the weather and climate but not for the sake of altering or controlling the seasons. They learnt about the world to understand and fit into it not to command or control it. (ibid.)

He identifies the 'mental models' or assumptions that persist in western school systems. Learning takes place in the mind not in the body as a whole, learning takes place in the classroom not in the world, knowledge is inherently fragmented. He argues for schools to both model and teach systems thinking.

The main features of systems thinking are relationships, interdependence, and diversity, which are three key ecological principles that sustainable schools will depend on and are core to the writing on social learning. Surfacing, testing, and improving pictures of the way the world works, including how the sustainability of the environment is excluded, is a critical part of its activity. Systems thinking is identified by Macgilchrist et al. (2004) as a key intelligence for a school. They define systemic intelligence as:

A way of thinking about the interrelationships and patterns that enable flow and connection between the parts that make up an organisational whole . . . it enables 'joined up thinking' and coherence. (142)

Systemic approaches see a school as a living system that is interdependent with the web of life of which it is a part, locally, globally, and environmentally (Capra 1996; 2002). In the event of climate change causing serious disruption to the patterns of conventional schooling a move to more community-based learning will be a critical step. Schools have to some extent defined their task in relation to their environment but this has largely been the way they think about the pupils and their social context once they are *inside* the school: less so the school's role in community building or environmental contribution *per se*.

The report *Every Child's Future Matters* published by the Sustainable Development Commission (SDC 2007) outlines the connections that need to be made between sustainability, the Every Child Matters (ECM) agenda, and social cohesion, connections that are providing educators with a hot topic at the moment. The networking and collective organization and sharing of resources to support health and wellbeing for families in the ECM policy agenda provides a possible blueprint for how groups of schools could attend more to ecological as well as social sustainability.

> We recognise the right of children to a safe, healthy, enjoyable and rewarding present—but are we acting to protect that quality of life in the future? Climate change stands in the way of this at present. It has the capacity to destabilise the economy here and overseas, producing upheaval, insecurity and poverty as well as incalculable environmental damage. Our challenge is to extend the horizons of children's policy beyond the present social and economic focus to embrace the environment as a key factor in wellbeing. (SDC 2007, 5)

Towards a Learning Orientation for Sustainability

As well as developing a more cohesive approach to education for sustainability through an ecological worldview, a more outward stance, and more systemic thinking, there is the question about whether school-improvement activities are as currently focused on learning and promoting the necessary learning orientation as is suggested by the social learning authors (Wals 2007). Hopkins (2008) identifies learning to be resilient as central to a global society facing climate change. He defines resilience as the capacity of a system to absorb disturbance and reorganize while undergoing change.

Concerned about a school system over-preoccupied with assessment outcomes, Watkins (2001) has drawn attention to the difference between performance and learning. His research indicates that an emphasis on performance can depress learning while an emphasis on learning can enhance performance. Bonnett (2004) suggests that we need to reconsider what counts as the learning and knowing appropriate to education for a more sustainable future and argues that many disciplines have their roots in a cultural paradigm that aims to conquer and exploit the natural world.

The main research and development work of the International Network for School Improvement at the Institute of Education, London University, that I have been leading with my colleagues for the past decade, has been promoting *Learning Focused* School Improvement (Lodge and Reed 2006). This is school improvement that is:

> designed to create a learning orientation for young people and is carried out simultaneously, not separately, at the level of the school, pupil and teacher with the underlying purpose of improving young people's capacity to learn. Learning focused school improvement is based on a more ecological and less mechanical model of how a school operates. (6)

Our research projects in partnership with local authorities around the country have been demonstrating for some time that learning is the weak relation in a performance culture and that teaching and attainment while depending on learning for their success are the more dominant partners. It is assumed in unexamined models of classroom pedagogy that 'good teaching' naturally leads to good learning. Social learning as defined by Wals (2007) assumes a strong role for the learners, that they can take responsibility for their learning, know how to learn, and know that their learning is a lifelong process that continues outside formal learning situations.

In talking to many groups of students around the country we have discovered that this is far from the case. Pupils across the age range in English classrooms are predominantly passive, think they need a teacher to be present to learn, that their teacher is the one responsible for their learning, and that their friends, the key to social learning, distract them rather than help them learn. They are imbued with the idea that learning and work are the same thing and that play and learning are two separate unrelated processes. A recent MORI poll commissioned by the Campaign for Learning (2008) has confirmed our research findings and that these trends are getting worse. The assumptions that underpin pupils' views and beliefs about learning in school are clearly part of the industrial model of schooling that has already been noted. Activity in the projects that we have run aims to engage the students more fully in the classroom and support them to see themselves as agents in their own learning. Their passive dependency on their teachers does not prepare them to take greater responsibility in a future world challenged by climate change. The teachers involved have reported greater interest in their professional work, greater connection to the students, and improved behavior and interest in classroom study by students.

Vare and Scott (2007) argue that for sustainability to happen it needs to be a learning process and not about a top–down approach to rolling out predetermined behaviors. This chapter has so far argued that the form of social learning needed for the whole-school system to face climate change is based on greater learning across the current divide between school improvement and ecocultural sustainability and that this is needed by all the stakeholders in partnership to design and test solutions.

HAS SCHOOL IMPROVEMENT ANYTHING TO OFFER IN A TIME OF INCREASINGLY DIRE SCENARIOS FOR CLIMATE CHANGE?

This final section is challenging to address. Why? Because it not only requires an imaginative leap into the territory that not many of us have yet dared tread, it also means facing the loss of hope that we could make it, unscathed, the hope that our children would not have to reap what we have sown and face the music of a planet heating up, the displacement and extinction of many humans and nonhumans on the front line, and the loss of livelihood and wellbeing as we have known it in the North for several centuries. It is unlikely that school improvement will continue as a discourse for very long, if it hasn't already been abandoned, for it depends for its identity on there being schools. From what we can see in the written extracts from the future that follow, formal education begins to take place in quite different settings and in different ways after about 2025.

The following three short accounts illustrate the kind of educational changes that may result from dire climate change and the resulting adaptations that would have to be made. They take place in London between 2025 and 2050.

The first is the voice of a senior educational adviser coming to the end of her career in School Restructuring for Ecocultural Sustainability (SRES) in 2035. She is speculating about the impact of the work she has done to try to ensure that schooling becames a more ecological enterprise:

> After all, school change and learning have always been an optimistic craft and they should have developed the capacity to save us from dire climate change. It's true, we have increasingly changed our school improvement direction in the past 25 years. We don't really call it school improvement anymore since the SRES initiative but the latter phases of school improvement certainly influenced what we do now. We have been learning more together with others in the field through developing lateral partnerships, giving school reform a better theoretical basis in ecology, teaching our young people about an ecological paradigm that connected them to the biosphere. Learning is much more of a social and community activity than it was when I did my PGCE in the 1990s. We finally behave as if learning takes place in relationship and in the body and the heart and not just the mind. As a result children and young people have become much better learners since the end of compulsory testing and exams at 16+ and are really up for what this is going to take. I just can't imagine a system with an emphasis on testing anymore, that period of educational history started to come to an end in the summer that the floods were so bad that schools across England hardly opened. That was a turning point for the school system, but none of it has actually saved us from the impacts that we are seeing now, none of it.

Second is the voice of a young woman in 2040 who is one of the lucky ones to have studied for a traditional degree course, given that the system has contracted substantially since the time when there were the resources to provide an academic qualification for anyone who wanted it. She was born in 2020 and is undertaking an undergraduate course in Education for Resilient Community Development. This is an excerpt from a chapter in the dissertation she is writing about the change of educational paradigm over the past fifty years. She describes the transition from the time when school improvement had been the dominant discourse about educational improvement to when the Northern hemisphere finally became more ecologically minded and realized wellbeing and quality of life should take precedence over standard of living and qualifications. She reflects on how formal learning has already changed in her lifetime:

> Life for my generation isn't bookish or learned like when my grandmother was at University at the turn of the century, it is essentially practical and about survival. We have evolved a way of educating our children and ourselves that is so different from what I am reading about educational change forty years ago. Of course children and young people need particular care, education and support. But for several years now since the floods happened regularly, pupils have been going to school only when they can open, no longer expecting to stay in desks or to have to listen to a teacher at front of the class. The schools are gradually closing especially the ones that hold the heat. We can't afford to keep them open all the time or to train enough teachers so children are beginning to go to school only for some of the time, and more and more parents are getting involved in going to and leading classes themselves and in learning with their children. School grounds that aren't in the flood plains are still the key place in the community for food growing which is the main thing that we all study and do together. This has changed how we think about formal education.

The third voice is of one of the elders in a community that has been developing the concept of Community Learning Groups (CLGs) since 2047. The CLGs took over from the schools when the system gradually collapsed in the 2040s. He is preparing to give a speech to open an end-of-year celebration to mark the middle of the century and is reflecting on the changes he has seen in the past fifty years.

> The biggest loss we have had in our generation is the hope that the planet herself could save us, as if that were ever a possibility. As the great teacher Thomas Berry said way back in 2006, to be educated is to know the story and the human role in the story (Berry 2006, 22). It is a great privilege to be alive at a time when it is dawning on the human community as it struggles to survive and faces the fate that its ancestors bestowed on it, what its place is in the whole. We are at last waking up to the indigenous knowledge our ancestors forgot for centuries. We are now alert to the complete dependence of our species on the life cycles and processes of the planet, over however many millennia it will take, to adjust to the ravages of climate change that have occurred and as a consequence be more limited and unpredictable in what it has to offer

humans. We are the great witnesses of this transition and are beginning to share and understand this knowledge better now through our CLG processes. The face of public education has changed irretrievably in my generation, as much as a result of circumstance as any intention of our own.

Yes, there are the remnants of the formal system, the older schoolhouses of the northern world still stand, but they are no longer used for the purpose that was intended for them. The buildings are strong and a beacon to an age that has gone. They are cool in summer and warm in winter. They have become community places to stay when it is too cold to stay in our own homes. They are where the community meets together to develop our skills and crafts, to study together to grow our food. With luck the resources, books and records from our great libraries and archives that we keep on the top floors are still going to remain accessible, they need as much as possible to be protected from the floods. So much has already been lost.

It's now too hot to concentrate for much of the year and when they are not helping with food growing or cycling on their bikes to look after the elderly in the care homes our youngsters are at the river or stream, or in the woods provided the flood warnings aren't on. Of course they learn to read and write and be numerate but learning and its development has taken on new meaning, away from the traditional schoolhouse.

My Aunt was a school improvement adviser in the 2020s. School improvement did much up to the point where we just didn't have the energy to keep the buildings open for young people's learning only. Young people have adjusted to a contracted situation and have long since been part of a culture that has matured and in the face of disruption and displacement and death is more able to face up to the transitory nature of its existence than our ancestors were. (Selby 2007)

CONCLUSION

Mulgan's (2007) work on social innovation shows that effective change results from new combinations of existing forms and practices. It also requires a change of heart. School improvement scholars question the assumptive base of the system, argue for more effective learning, and cite theory from the field of quantum theory, the new physics, and current organizational discourse (Mitchell and Sackney 2000; Stoll et al. 2003; Fullan 2005; Sergiovanni 2005) but without a worldview that is strongly about ecological sustainability. Colleagues in the sustainability field have strong constructions of social learning that assume a purpose of sustainability but don't necessarily concern themselves with the implications of the changes they are advocating or the outdated views of learning that are blocking progress. The NCSL Sustainable Schools project provides a hopeful way forward for the two fields to inform each other more explicitly. It shows a group of schools in transition from one paradigm to another, from the industrial to the ecocultural model. The signs from the project are at this early stage of evaluation that head teachers are well informed about sustainability, know their students are growing up in an uncertain future, and that their task is to empower them in practical solutions and more participatory approaches. They are at the foothills of pioneering new ways of learning together, new content form, and organizational arrangements for those who are to follow.

The question still remains: as Brown (2008) says, is there enough time for others in the system not as yet alert to the urgency of the agenda to make the same commitments and changes or are we just tinkering our way into a transition that makes the dire scenarios of climate change inevitable? The insights into the three extracts from future times would suggest that time is of the essence. Yet it is worth resourcing ourselves with Macy's words (2005):

> Work on climate change is not served by playing upon our fear or guilt but by an enthusiastic release of creativity and ingenuity such as happens when we feel ourselves called on a great adventure. (5)

She encourages us to begin to see this work as "The Great Turning", a move from a life-denying to a life-enhancing society, and in which school change, learning, and transformation need to play a crucial role. The possibility of learning to change, however, still hangs in the balance.

SNAPSHOTS OF SOCIAL AND TRANSFORMATIVE LEARNING

I held discussions with two colleagues working in the field of sustainability in different ways to raise awareness of climate change. I asked if they would share their views. The first scenario takes place now, before a climate apocalypse and the second one after it. The question put to them was: what are/would students and teachers be doing to prevent or manage the worst effects of climate change?

Scenario 1

We have set up a project in the Calderdale area that involves children and teachers working together on practical solutions. They learn how to do a carbon audit including how everyone travels to school, the energy systems of the school, and how to improve them. This then involves head teachers in seeing ways forward and using the leverage they possess to change things. It's about consciousness raising and practices, not too much information.

The other important issue is to work on what type of schools are wanted now and to use a participatory frame to move from aspiration to how to get there. The 'Building Schools for the Future' of the DCSF hasn't got this orientation. They need to imagine gardens, growing areas, tree plantations. We need to bring a community perspective more into schools, develop the land they have, and grow food with the students so it becomes part of every-day life. Schools in Halifax have all taken the boats that were discarded from the municipal pond for seed growing! Most people are freaked and disempowered if you talk about the connection to climate change. We need to build resilience; for example, in Todmorden we have worked out it will take 212 acres for us as a whole community to be self-sufficient. We need to renew the rural/urban links and see them in new relationship. We need 2,200 eggs a day for our town. How do we source them locally? Everyone is publicly declaring the number of chickens they have and working at ways of making us sustainable locally. Bits of the curriculum in school need to be freed up to soil, digging, composting, manure. Thirty percent of species will be gone in thirty years. If we didn't have a predetermined curriculum we would start with that issue! Having a voice and constructing things together. Head teachers are still consumed with the performance outcomes of their school and we have ten years if we are lucky to get them all to real-ize we have to teach for a post oil/unstable climate agenda.

Scenario 2

The context is continual fluctuations in weather, flood protection sys-tems have gone; what we are learning isn't that different, how the world works, interdependence, and systemic awareness. We are thinking how did we get into this mess and noticing that the paradigm shift that had started hasn't saved us after all. A lot of what we are teaching is social skills. These are based around love, and compassion and care and are the basis for the resilience that we have to develop. Survival is the name of the game: to grow, use the little energy that we have. We are learning to live with fluctuation, change, and death. We have moved to an era of mass localization, stories we tell are of the travels that our grandparents used to make. Support networks are very important and there is a changed dynamic of how we get around and whom we see, as we have to cycle everywhere most of the time. The education system is based in the local community. Half of my week I am growing vegetables and in the other half contributing to community building. Children come to grow vegeta-bles with me and their teacher is putting that in context for them linking it to the science and knowledge of food. So teaching is much more distrib-uted and localized. There is sharing across places and communities. What does The National Learning Group do? They travel around and support our efforts to educate each other and they put us in touch with the efforts of our neighboring communities.

NOTES

1. A growing movement in the United Kingdom to create positive, localized community development and sufficiency and plan for less energy to be used.
2. This diagram was used in the presentation of the NCSL research findings in 2007.
3. A way to look beyond political boundaries to examine how ecological dynamics unite regions. Bioregions rarely overlap with counties, states, or countries—most bioregions transcend these artificial lines. http://www.permatopia.com/dictionary.html.

REFERENCES

Beare, H. 2006. *How we imagine schooling in the 21st Century.* London: Specialist Schools and Academies Trust.
Berry, T. 2006. *Evening thoughts.* San Francisco: Sierra Books.
Bonnett, M. 2004. *Retrieving nature.* Oxford: Blackwell.
Bowers, C. A. 2001. *Educating for eco-justice and community.* Athens, GA: University of Georgia Press.
Brown, L. 2008. *Plan B 3.0.* New York: Norton.
Campaign for Learning. 2008. *State of the nation survey 2008.* London: Campaign for Learning.
Capra, F. 1996. *The web of life.* London: HarperCollins.
———. 2002. *The hidden connections.* London: HarperCollins.
Clarke, P. 2000. *Learning schools learning systems.* London: Continuum.
———. 2009. Forthcoming. *Sustainability and improvement: A problem 'of' education and 'for' education improving schools 12.*
Cuban, L. 1995. The myth of failed school reform. *Education Week,* November 1.
Department for Education and Skills (DFES). 2005. *Learning for the Future.* London: DFES.
Forum for the Future/UCAS. 2007. *The future leaders survey: 2006/07.* London Forum for the Future/UCAS.
Fox, M. 2006. *The A.W.E. Project.* Ontario: Woodlake.
Fullan, M. 2005. *Leadership and sustainability: System thinkers in action.* London: Sage.
Glasser, H. 2007. Minding the gap in social learning. In *Social learning towards a sustainable world,* ed. A. Wals, 33–62 Wageningen, The Netherlands: Wageningen Academic.
Griffin, S. 1996. *The eros of everyday life.* New York: Knopf.
Hargreaves, A. and D. Fink. 2005. *Sustainable leadership.* San Francisco: Jossey Bass.
Hopkins, D. 1987. *Improving the quality of schooling.* Lewes, DE: Falmer.
Hopkins, R. 2008. *The transition handbook.* Dartington, UK: Green Books.
Lodge, C. 2008. *Student voice and learning—focused school improvement, INSI Research Matters series 32.* London: Institute of Education, London University.
Lodge, C. and J. Reed. 2003. Transforming school improvement now and for the future. *Journal of Educational Change* 4: 45–62.
Lodge, C. and J. Reed. 2006. *Towards learning-focused school improvement, INSI Research Matters series 28.* London: Institute of Education, London University.

Macgilchrist, B., K. Myers, and J. Reed. 2004. *The intelligent school*. London: Sage.

Macy, J. 2005. *Facing climate change and other great adventures*. http://www. COINet.org.uk.

Mitchell, C. and L. Sackney. 2000. *Profound improvement*. Lisse, The Netherlands: Swets and Zeitlinger.

Mulgan, G. 2007. *Social innnovation*. London: Young Foundation.

National College for School Leadership (NCSL). 2007. *Leading sustainable schools*. Nottingham, UK: NCSL.

Office of Standards in Education. 2008. *Schools and sustainability: A climate for change?* London: OFSTED.

Orr, D. 1993. *Ecological Literacy*. Albany: State University of New York Press.

———. 2008. The current context of climate change presentation. Presentation to the Sustainability Special Interest Group meeting. London: Institute of Education London University.

O'Sullivan, E. 2001. *Transformative learning*. London: Zed Books.

Reed, J. 2008. Can education change its climate in time? *Professional Development Today* 11(1): 17–20.

Sackney, L. and C. Mitchell. 2000. *Profound improvement*. Lisse, The Netherlands: Swets and Zeitlinger.

Sackney, L. and K. Walker. 2006. Leadership for knowledge communities. Paper presented to the *Annual Conference of Commonwealth Council for Educational Administration and Management*. Nicosia, Cyprus.

SDC (Sustainable Development Commission). 2007. *Every child's future matters*. London: SDC.

Selby, D. 2006. The firm and shaky ground of education for sustainable development. *Journal of Geography in Higher Education* 30 (2): 351–65.

———. 2007. As the heating happens: Education for sustainable development or education for sustainable contraction? *International Journal of Innovation and Sustainable Development* 2 (3/4): 249–67.

Senge, P. 2000. *Schools that learn*. London: Nicholas Brearley.

———. 2006. *Learning for sustainability*. Cambridge, MA: SoL.

Sergiovanni, T. 2005. *Strengthening the heartbeat*. San Francisco: Jossey Bass.

Specialist Schools and Academies Trust (SSAT). 2007. *Raising standards: Making sense of the sustainable schools agenda*. London: SSAT.

Sterling, S. 2001. *Sustainable education*. Dartington, UK: Green Books.

Stoll, L., D. Fink, and L. Earl. 2003. *It's about learning*. London: Routledge-Falmer.

Tilbury, D. 2008. Learning based change for sustainability: Perspectives and pathways. In *Social learning towards a sustainable world*, ed. A. Wals, 117–33. Wageningen, The Netherlands: Wageningen Academic Publishing.

Van Velsen, W. G., M. B. Miles, M. Ekholm, U. Hameyer, and D. Robin. 1985. *Making school improvement work*. Leuven, Belgium: Acco.

Vare, P. and W. Scott. 2007. Learning for change. *Journal of Education for Sustainable Development* 1 (2): 191–98.

Wals, A., ed. 2007. *Social learning towards a sustainable world*. Wageningen, The Netherlands: Wageningen Academic.

Wals, A. and T. van der Leij. 2007. Introduction. In *Social learning towards a sustainable world*, ed. W. Arjen, 17–32. Wageningen, The Netherlands: Wageningen Academic.

Watkins, C. 2001. *Learning about learning enhances performance, NSIN Research Matters series 13*. London: Institute of Education, London University.

Wrigley, T. 2003. *Schools of hope*. Stoke on Trent, UK: Trentham.

9 Critique, Create, and Act

Environmental Adult and Social Movement Learning in an Era of Climate Change

Darlene E. Clover and Budd L. Hall

INTRODUCTION

Awareness of the need for climate change adaptation is growing as the Inter-governmental Panel on Climate Change (IPCC 2007) projections for weather extremes are validated around the globe. The pressure to develop responses is also increasing, as it becomes clear that global emissions are now higher than the concentrations on which the IPCC's worst-case calculations are based, and therefore climate change impacts may occur faster and be more severe than originally thought (Global Carbon Project 2008). Climate change efforts focus broadly on strategies of adaptation or mitigation or some combination of the two. Efforts to curb the production of greenhouse gases by, for example, polluting industries are aimed at mitigating the worst impacts. Given the evidence however that climate change is already upon us and may at best be slowed somewhat, more and more of the focus is on adaptation strategies.

A major conceptual factor behind climate change is the global economic system. Kumar suggests the "economy is a wholly owned subsidiary of ecology" (2007, 3). And yet the paradigm sweeping the globe is one where the economy comes first, the environment its handmaiden, and growth and development its voracious practice. Moreover, scientists, strategists, activists, and even some politicians argue this economy-first paradigm lies at the heart of climate change—the concrete outcome of global capitalist development that ignores people and ecosystems. Citizens need to be mobilized since this "environmental reality is linked powerfully with other realities, including growing social inequality and neglect and the erosion of demo-cratic governance and popular control" (Speth 2008, xi). Others, however, contest this viewpoint and they often have better access to government support and mainstream media. For example, Margaret Wente's column in the *Globe and Mail* (2009), Canada's only 'balanced'—relatively speak-ing—national newspaper, identified 1998 as the hottest year on record and therefore noted that the planet was not in fact heating as predicted. Scien-tific reports of melting of polar ice caps, turbulent weather patterns, grow-ing food insecurity, and dwindling natural resources are not the outcome of unfettered growth development but are something more 'natural.' The message is that citizens should make small changes in their lives and governments will look after the larger and long-term issues in their interests. As expected public confusion, fear, and even sometimes apathy are often the result of these inconsistencies in messages (Jickling 2001).

Contesting viewpoints and paradigms in this age of climate change are also found in educational responses and approaches. Some educators choose to take up the individual change and behavior modification approach as supported by governments and even organizations such as UNESCO. They focus on educating individuals to make personal and incremental changes in their lives. Their practice revolves predominantly around disseminating lists of activities people can do to lessen their impact. Activities focus on calculating personal CO^2 or greenhouse gas contributions, recycling, home refitting, and so on. Others, however, resolve to situate their educational work within a broader political framework. Learning in interesting times calls for interesting socioenvironmental education practices that include critique, creativity, and active public engagement that both contest and exercise power in this contemporary landscape. Individual changes are fine but mobilizing against corporate and government activities that contribute far more to climate change is imperative.

This chapter tells the story of one example of environmental adult and social movement learning: a critical, creative, collective process on Vancouver Island, Canada, around the proposed construction of a corporate-government, Canada-United States gas-fired power plant. Deemed "The Positive Energy Quilts: A Visual Protest" this arts-based environmental and social movement learning project was undertaken by a group of women artists and activists to oppose construction of the new pipeline and power plant. With needle and thread, these women created an important space to unravel consent and help people speak out creatively and collectively as concerned citizens. We argue that the strength of this work is the privileging of people's knowledge over expert-driven knowledge, the encouragement of collective creativity, voice, and agency through an arts-based participatory process and gentle but persistent public acts of defiance and dialog that helped to defeat a major development project. In telling this story, we do recognize that winning is often not only the result of community-based education and activist struggles. Sometimes the changes happen many years later as some of the same people who have learned from today's struggles continue to work on into the future. Our aim with this example, however, is to show the power and potential of a climate change–oriented community-engaged educational process and how "if we come together . . . there is hope" (Reimer, forthcoming).

ENVIRONMENTAL ADULT EDUCATION

Before we move to the story of the educative-activist activities around the gas-fired power plant on Vancouver Island, it is important to situate it within a discussion of environmental adult education and social movement learning.

Environmental adult education represents a movement away from expert information–driven paradigm of much public environmental education towards the idea of political empowerment, people's knowledge, and collective action for change. While mainstream public environmental education often focuses on information transmission, environmental adult education works as an engaged and participatory process that begins with recognizing and respecting people's knowledge and bringing these together through dialog and debate to create new ecological understandings of our world. In terms of climate change, environmental adult educators do realize that many people may not fully understand the complicated science involved. However, this does not mean they are without ecological knowledge that can contribute greatly to debates and discussions. Having said this, always within a group of adults in a community there will be someone such as the local clock repairman who will have a deep understanding of the science and be able to add this knowledge.

Environmental adult education also moves away from the individual change, behavioral modification paradigm. This does not mean that individual change is not valid but simply means that to bring about profound change quickly we need processes of learning and engagement that draw attention to the politics behind corporate and governmental activities that contribute to climate change by privileging economic growth and development. Environmental adult education calls for critical and creative debate and actions that can redemocratize climate change debates by placing people and the ecosystem at the centre. As corporations grow stronger and governments move away from the environmental agenda at lightening speed in any time of crisis, we need educational processes—both formal and nonformal—that dramatically strengthen participatory democracy, collective voice and energy to work for change (i.e., Clover and Hill 2003; Hill 2006).

In terms of practice, environmental adult educators whether in formal learning institutions such as universities or working nonformally in the community use a variety of practices, strategies, and tools in the praxis of learning to encourage critical thinking, active debate, and, most importantly, creativity (Clover et al. 2000). These practices actively encourage people to reimagine and reshape the world as they wish it could be. They often include learning circles and small group discussions, but more and more adult educators have begun to focus on the aesthetic. Collective mural and music making, popular theatre, fabric arts, or a combination of these are growing in use and popularity. Throughout the world, art-based environmental learning is used to bring diverse sectors together (such as unions and environmentalists), give a new creative voice to people's environmental concerns, transform consciousness, and challenge processes of development that embattle the planet (see Branagan 2005; Clover and

Markle 2003; Keough et al. 1995). This shift is happening due to a recognition that the arts are a medium through which notions of self and community can be both experienced and expressed. They enable people to cross the empty spaces between themselves and others. The arts are valuable tools of connection to help build new partnerships and relationships across race and other practices of difference. Of all our cognitive capacities, the imagination is the one that permits us to give credence to alternative realities. It allows us to break from the already assumed, the taken for granted, the confines of privatism, and set aside these distinctions and definitions that separate and divide (Greene 1995).

Social movement learning refers to: (a) learning by persons who are part of any social movement; and (b) learning by persons outside of a social movement as a result of the actions taken or simply by the existence of social movements (Hall and Clover 2005; Hall 2006). What we all know as facilitators of learning is that nothing is as powerful a stimulus to learning as the necessity to teach or inform others. The organizational or communicative mandate of all social movements is a necessarily educational concern. There are two main types of learning within social movements. One is nonformal. This includes organized or intentional educational activities organized within the movement itself. The Land Conservancy (TLC) in Victoria, British Columbia, Canada, is a local nonprofit organization and part of the larger movement towards adaptation or mitigation of climate change. On their Web site they note, "One of TLC's fundamental purposes is to provide education to the public about the importance of conservation and the protection of our natural and cultural heritage." If one further explores their educational outreach program, one sees that they engage in: site interpretation, school visits and talks, a speaker's bureau, and cottages for study groups. Likewise Greenpeace UK builds public education into its campaigns. It also supports study groups through its list of experienced speakers. Both of these groups use the Internet as a vast education environment, an open-source pedagogical design location that combines nonformal educational initiatives with social networking strategies, Web-based petitions, and fundraising.

A second way in which social movement learning occurs is informal or incidental in nature. The bulk of learning by those within any movement is of this nature. Whether one is working on a local climate change adaptation or mitigation issue or on a global campaign, if one is responsible for designing a Web site, creating informational materials, writing a petition, finding out what policies exist that are inhibiting change, one is forced to learn both the depth of the climate change issue and the strategic points simply to be able to do the job within the movement. All persons in climate change movements are learning at an incredible rate and learning more than they may well have anticipated.

There is still another powerful form of social movement learning and one often neglected in the literature. It is the learning that takes place by persons who are not directly participating as members of a given social movement, by people outside of a given movement. All of us, no matter whether we are formally informed of the intricacies of climate change or not have been learning about climate change because of the nearly thirty years of work done by activists and movements. The actions of social movements, be they large-scale media events that such as Greenpeace and other environmental groups have staged, or local land occupations or even bird-counts to document the loss of biodiversity as a result of climate change, create rich environments for learning by large numbers of the public. The educational instruments used by social movements are varied and creative: Web-based social networking strategies, poetry nights, marches of all descriptions, large-scale puppetry, radical cheer leading, collective construction of art works, popular and forum theater, panel discussions, song writing and singing, newspaper articles, and other forms expressly created to carry the messages of the movements beyond the boundaries of movement membership. Such is the power of social movements to reframe the world, that none of us escapes this ongoing democratic flow of energy. To this transformation of vision and imagination, we need to give the name learning (Transformative Learning Centre 1994; Hall 2004).

VANCOUVER ISLAND DEVELOPMENT: THE CONTEXT

Often called Canada's subtropical coast, Vancouver Island is considered ripe for economic growth and development. With its British Isles moderate and temperate rainforest climate, numerous sandy beaches and wilderness parks, year-round fishing, surfing, and diving, picturesque villages and coves, and numerous golf courses, Vancouver Island has become a magnet for people from across Canada and around the world. With doors thrown wide open to unfettered development by the current neoconservative government of the province, many jump onto the bandwagon. Private developers find fertile ground to profit from the influx of new home buyers and create elaborate plans for new subdivisions, for example, that sprawl across the landscape, eating up green space (Clover and Markle 2003).

With an increase in population comes an increase in energy needs and the opportunity for expansion. The once fully publicly-owned utility BC Hydro is actively developing stronger partnerships with energy corporations in the United States to expand power generation. One plan, and the platform from which our story springs, was to construct a natural gas–burning power plant at Duke Point, Nanaimo, Vancouver Island. There are, as one would expect, contesting views around this source of energy. On their Web site, the International Petroleum Environmental Conservation Association (IPIECA) argues that "natural gas has the potential to play a significant role in a carbon-constrained energy future as a

relatively low-carbon fuel source." The David Suzuki Foundation coun-
ters that although natural gas has some advantages over other energy
sources, relying on natural gas is still problematic. Gas-fired plants do
emit sulphur dioxide and nitrogen oxides that contribute to acid rain and
ground-level ozone, both of which can damage forests and agricultural
crops.

BC Hydro's idea was to include a new pipeline that would carry natural
gas from Sumas, Washington, to a new gas-fired power plant proposed to
be built on a sensitive ecological location. The purpose of the power plant
would have been to generate surplus electrical power that could be sold to
the United States while creating some jobs for an area that was in need of
new employment strategies given the major downturn in many traditional
areas such as mining and forestry. The basic rationale, as quoted on the
Web site Nanaimo Citizens Organising Committee (NCOC) in 2003 was
that "with growing North American demand, especially for natural gas
used in power plants, [we must] support continental energy security."
However, on the other side, the perceived socioenvironmental risks associated with the proposed Duke Point Power Project (DPPP) ignited a fire
under many citizens not only on Vancouver Island but the smaller islands
just offshore as well. Almost immediately, a diverse coalition of environmentalists, activists, artists, and a variety of others began a series of organized, nonviolent oppositional activities to the plant (Mace 2005).

Before construction could begin hearings and public meetings were
mandated by the government. These were problematic for a number of
reasons. First, developers disseminated artistic renderings—often ideal-
ized computer graphics—which did not tell the whole story of the environmental impact. The reams of data and information produced by BC
Hydro focused on the economic benefits and pending blackouts and
brownouts the Island would experience without the project. This polar-
ized a community desperately in need of jobs on the one hand but also
very environmentally conscious on the other. Second, while these public
meetings purported to 'give voice' to community they were seen merely
as a facade of consultation since many felt the decision to develop had
already been made and the corporation was simply going through mandated motions as is so often the case in Canada. Third, and related to
the last, the inequitable power dynamics that surrounded this plant and
the public information gatherings discouraged and silenced adversaries.
Although there are many reasons for this, a key factor is that the correct
and most important information and knowledge about the project was
seen to lie with the corporation, rather than the people who made up the
community. In other words, as Saul (1997) notes, the process was simply
a cloud of information that emanated largely from the corporation and
the government as rhetoric which obscured their agenda and power and
acted solely "as a steam-release device for the general public" (in Branagan 2005, 37).

THE STORY

Darlene Mace (2005), a local adult educator and artist who lives on one of the islands that would have been affected by the gas-fired plant, carried out a research project that documented the learning experiences of some of the participants in the movement for a Master's thesis titled: *A Change of Mind: Social Movement Learning and the Cancellation of the Duke Point Gas Plant (Vancouver Island and Gabriola Island 2005)*. Mace's work illustrates a number of pedagogical moments in the DPPP movement that help us understand the importance of environmental adult learning within social movements and provide some ideas for all of us as we consider the challenges of climate change within a world of capitalist development and growth. Perhaps the most fascinating example—one in which Mace was a part—of environmental adult education and social movement learning around the proposed DPPP was a collective quilt project—The Positive Energy Quilts: A Visual Protest—undertaken by a group of women artists and activists determined to engage the public and engage a broader different voice and aesthetic form of activism.

In response to the proposed power plant, approximately 400 people attended the first public meeting sponsored by the Nanaimo Citizens Organising Committee to discuss ways to develop a more informed public and strategies for opposition. Fabric artist Kristin Miller attended this meeting because, as she noted, "This power plant thing was in my front yard." During the meeting, Kristen agreed to join the Canvassing Committee and engage in outreach and educational work. Since Kristin had never before in her life been involved in anything political, she spent the first part of the meeting wondering how she could get out of this when suddenly an idea struck her which would involve her skill in fabric craft—creating something that would be a form of protest and voice. She argued to the group that women were used to tackling problems with needle and thread so it was a logical thing to do. She explained the quilts could be a medium through which a community could express its thoughts and feelings about the proposed plant. "And they all just looked at me like I was crazy—at first."

Over fifty people immediately contacted Kristen to say they wanted to work on the quilt project. Many had never quilted before; others had never been politically active. But what they all had in common was a desire to connect with the visual protest. A core group of fifteen women artists and activists was struck who disseminated squares of fabric to elder-care facilities, artists and community organizations, schools, and to a number of other venues. No guidelines were given for what could/should go onto the quilt other than the suggestion that people make images or write words that expressed their feelings and reactions to the proposed power plant. There was an assumption, as Kristin remarked, "that [you] were meant to show opposition to the plant and so if you were in favor you maybe did not get what we were talking about!" Within a relatively short period, over forty squares were returned. Some were created by an individual, some by a group but all were colorful and heartfelt, ranging from very traditional quilt patterns to more strongly political images and ideas.

The core group of fifteen women came together to arrange and rearrange the squares until they finally reached consensus on the layout and then the sewing began. In order to make a stronger educational connection, the women decided to engage in what they referred to as 'quilting in public.' They began by quilting outside an arts shop and café on Gabriola Island. The women then decided it was time to quilt in public at a rally opposing the gas plant. This was followed by a quilting bee beside the local TV studio and police station. As Kristin remarked, "That was the first time the police noticed us." Together, the women and the police discussed the legality of quilting in public—and here you will be delighted to know that quilting in public is in fact legal in Canada unless it is attempted on an aeroplane and then, of course, the needles would immediately become weapons of mass destruction—and the women were left to quilt in peace and to show off their quilts to the TV cameras. Following this triumph, the group decided to do something even more controversial and set up their quilting operation

Figure 9.1   The quilts were made in 2002, to protest the construction of a gas-fired power plant on Vancouver Island, at Duke Point near Nanaimo, B.C.

on the sidewalk outside the building where BC Hydro was holding a public information session. Numerous curious passersby stopped to chat with the women about what they were doing, and discuss the issue.

Once the quilts were completely finished, they were taken for display at the interveners table of the Environmental Assessment Hearings about the gas pipeline that would be linked to the plant. However, upon their arrival, the administration at the site said they were "too political" and they were not allowed to be put on display. So the women moved them to an active community center nearby. The quilters then again tried to take the quilts into the Utilities Commission Hearings but were asked to leave, again for the same reason. So the group stood with their quilts outside on the sidewalk and again drew the attention of people walking by and those leaving the hearings. The quilts were worn like cloaks at many other public events such as a protest march with hundreds of others who followed a brass band up the hill to City Hall. The quilters slipped quietly into council chambers and hung the quilts on the walls.

Some people who made the squares took a softer approach, stitching windmills, shrimps and scallops, or solar panels to identify various aspects of the environment that would be harmed or draw attention to alternate energy forms. Others connected economic gain with environmental plundering such as the image of a stream of satiny water disappearing down the throat of a coin purse accompanied by the caption moniker www.stupidity. com. In the same vein, another square read, "What a bleak future, if we plan with a short-sighted vision of wealth and a barracuda mindset." Still other participants in the quilting activity showed a deep understanding of the polluting and toxic components of the plant. For example, the image of a red devil in the shape of oil gushing forth (poetic license to which we return in a moment) is surrounded by the words "ammonia," "carbon monoxide," "volatile organic compounds," and "particulates."

A number of squares made bolder, more political statements. On one sits a comical image of Uncle Sam roasting the world over livid orange flames spewing from a power plant. A second is the stripes and stars found on the US flag, with a maple leaf replacing one star and carrying the words "No, eh." A final image is of Uncle Sam making a rude gesture, with the words "Uncle Sam wants YOU . . . to build his power plant."

Returning to the issue of poetic license touched on above, one group created a square with the image of a factory spewing thick gray-black smoke. At first, the core group of quilters attempted to argue with them because they knew BC Hydro had said only white steam would be emitted. It is important to note here that in the world of activism, as many of you know, there is always the issue of getting the environmental facts right if one wishes to be taken seriously by those in the 'know.' However, the group that produced the square argued that the smoke was gray because it represented dirt. They believed that what was coming out of the stacks would still be polluting, and they did "not want to breathe it, even if [we] can't see it." Through much debate, the group came to see the gray-black smoke as symbolic of the pollutants, trace elements, and particulates that BC Hydro admitted would be emitted in the steam. The sooty color black smoke may also symbolize the quilter's own feelings about an industrial plant in her backyard, or even dismay at the global climatic effects of burning fossil fuel to produce electricity.

POWER AND POTENTIAL

Moore Lappé (2009) questions what keeps us creating in a world that often violates our deepest values in a downward spiral towards climate chaos. Although The Positive Energy Quilts activity was only one of a number of protest and outreach actions taken against the plant, it was extremely effective in helping to turn the tide against the project. We conclude this chapter by highlighting, through the lenses of environmental adult education and social movement learning, some of the interwoven aspects of this project that made it so powerful.

Respecting Knowledge

Corporations and governments alike have the resources, the time, and the human-power to mobilize the data and information required to launch a full-scale assault on communities that do not have such resources at their disposal. Moreover they can quickly organize public consultations and produce drawings, glossy pamphlets, and charts and disseminate these far and wide. They are produced often under the guise of helping people to make 'the right' decision but their aim is often to seduce and coerce. Something seldom called for is the knowledge of so called ordinary people—what they know, think, and feel about the issues in ways which are truly expressive. Recognizing, tapping into, and validating people's ecological knowledge is a foundation of environmental adult education and was a foundation of this project. People were asked through symbol, color, imagery, and text to speak out and share their knowledge, which ranged from understandings of the toxic by-products of the plant to the larger political situation.

Public Engagement

By using the process they did, the quilters were able to engage the public in two very important ways. The first was to reach out, through the squares of fabric, to people around the community who had deep concerns about the proposed power plant but were not necessarily the type—or of an age, such as seniors in care facilities—to participate in rallies or attend public hearings. Just because people are not or cannot be active in the more traditional ways does not mean they should be excluded. In fact, they were very much present, their issues and concerns very visibly stitched into the mosaic of the quilt. Picking up on this stitch so to speak, by quilting in public in so many diverse locations, the quilters were able to engage many other people in process. Kristen in particular believed perhaps it was the soft, gentle approach as well as the beauty and color of the quilts that drew people in and then they gradually began to get the message that a serious concern was being addressed. In fact, one quilter said:

> One man came up and he was almost in tears and he said, "How can I find out about this?" He . . . was struck by the art and he hadn't known anything about what we were protesting. And it didn't matter what else was going on in the street, this man had to talk to me. He said, "I don't even know which questions to ask" . . . And he was really feeling devastated in himself, so we had quite a good conversation.

The Arts and Attention

One thing that will always be certain—where there is art there will be censorship. It is interesting to note in the story that it seems to be totally acceptable to invite community members to a public hearing to voice their concerns but it does not appear to be acceptable to have a concrete visual of those concerns hanging on a wall at the public hearing. After all, it was the quilts that were tossed out of the public forum, not the audience. This is because, as Griffiths (1997, 30) suggests, the arts are understood to be "far more than mere self-expression or decorative pastime." The counter-images projected on these quilts combining beauty and politics, facts and feelings, raised red flags for some, who believed they needed to be stamped out because they were so "vigorously effective" (31). The arts are attention-grabbers and attention is something that can be useful to social movement actors aiming to get their point of view out to the larger public. The quilt-ing-in-public activity and the quilts themselves attracted hundreds of pass-ersby who stopped to enquire, listen, and discuss the issues. Some professed they had not heard what was going on and were pleased to listen to what the women had to say. In addition, the arts are a darling of the media. The women were photographed and interviewed often, something that would not have happened to the same degree had they simply sat outside the hear-ings handing out protest leaflets. It was the quilts, these soft, innocent yet subversive and highly political objects that drew the media over and over.

Visual Counter Stories

Stories are what we use to grab things, to hold onto them. Wherever the quilts were put on display, people were asked to 'read' the messages they contained. The quilts told a story of a community—a past, a present, and a future. The images and stories reconceptualized, deconstructed and undermined the 'promise' of development and created new conceptual and analytical spaces of community landscape and life. In addition, the power plant officials were extremely confident with their charts and graphs, so convincing with their arguments and stories of employment or the prevention of blackouts, and so logical that it was virtually impossible for people to fight back using solely the same tactic. As Kristen noted, "We had strong feelings, but they were not necessarily logical, and the facts, other than those pre-packaged by the power companies, were surprisingly hard to find. I could not argue at their level but I could say what I thought through my quilt square and encourage others to do the same."

CONCLUSION

IPIECA notes astutely that although natural gas technology today can dramatically increase supply to energy markets, its application must "overcome an array of . . . environmental and social barriers before its full potential can be realized." One of these barriers is clearly an active, vocal, and creative public. The story of the positive energy quilts is an example of the power of contemporary environmental adult and social movement learning as a critical component of the criticism, creation, and action required to mobilize and support citizens in moving an agenda of sustainability and social justice forward in today's complex times. No matter how solid an evidentiary case made by the scientific community for action on climate change, the political will to act will not be generated in the absence of broader citizen engagement and demands. We conclude this chapter with two scenarios of our own creation—of what the future world will look like with and without environmental adult education and social movement learning. A third scenario would involve the coexistence of the two scenarios.

Scenario 1

It is the year 2099, the dawn of a new century. To say things are not going well is an understatement. Oil, coal, and natural gas remain the primary sources of energy although we are into the final reserves that are now running dangerously low. We have managed to make them stretch this long because only a very few countries have access to them. The air is clogged with nitrous oxide in the only countries—the Western countries—that still uses these fuels. But they are on the brink of extinction. Citizens have recycled, avoided airplane travel as much as possible, and many long ago moved to public transport. In spite of the hopes of the early twenty-first century when the Intergovernmental Panel on Climate Change issued its reports and when both the IPCC and Al Gore were given Nobel Prizes for their work on climate change and sustainable development, the combined power of global capitalism linked to arms manufacture and the thrashing of the

American empire in decline managed, in the absence of a broad-based and active citizen involvement, to escalate political processes from global cooperation to global competition. Violence and the struggle for the resources of Africa, the Arctic, the Antarctic, and other once pristine areas of the earth have brought humanity to the brink. Life expectancies, which had grown steadily throughout the twentieth and early twenty-first centuries, have declined. The wealthiest five percent of the world live in luxury unheard of in earlier times, protected for the time being by high-tech security of every description. People have given up organizing environmental adult education workshops and other social movement strategies for change. Hundreds of their people are lying and dying in jails charged under now broad terrorist legislation.

Scenario 2

In 2010, fuelled by a global economic crisis and a realization that global capitalism has stumbled and fallen, smaller and larger groups of people in their local communities, flying the flags of anarchism and earth-based spiritual traditions, linked with green builders and green business leaders, come together, not in a formal organization, but in a spirit of hope, of change, and a profound belief that the human capacity *to learn* offers strong support to the reenchantment of the earth. The IPCC recommends that the only way to generate the deep engagement by people of the world is by going beyond a faith in scientific fact as the solution. They create community-university research partnerships respecting the knowledge(s) of the ecological traditions of the earth, provide funding for study circles, meals of baked bread, alternative transportation schemes, Web-based and face-to-face opportunities for all of us to create the knowledge that we need to transform ourselves and our communities in the direction that we universally aspire towards. No science degree in the world is awarded without attention to the learning implications for that science. Movements become part of the teaching and learning within our formal educational institutions. We learn from the poorest and most marginal at last. By 2099, we are closer to the world we want than at any other point in planetary history.

REFERENCES

Branagan, M. 2005. Environmental adult education, activism and the arts. *Convergence* 38 (4): 33–50.

Clover, D. E. and G. Markle. 2003. Feminist arts practices of popular education: Imagination, counter-narratives and activism on Vancouver Island and Gabriola Island. *The New Zealand Journal of Adult Learning* 31 (2): 36–52.

Clover, D. E. and L. Hill. 2003. *Environmental adult education: Ecological learning, theory and practice for socio-environmental change.* San Francisco: Jossey-Bass.

Clover, D. E., B. L. Hall, and S. Follen. 2000. *The nature of transformation: Environmental adult and popular education*. Toronto: Transformative Learning Centre, OISE/UT, and International Council for Adult Education.

David Suzuki Foundation. http://www.davidsuzuki.org/Climate_Change/Energy/Fossilfuels/naturalgas.asp.

Global Carbon Project. 2008. Carbon budget and trends 2007. http://www.global-carbonproject.org/carbontrends/index.htm . Toronto, ON. http://www.iclr.org/research/publications_climate.htm (accessed February 8, 2009).

Greene, M. 1995. *Releasing the imagination*. San Francisco: Jossey-Bass.

Greenpeace UK. http://www.greenpeace.org.uk.

Griffiths, J. 1997. Art as a weapon of protest. *Resurgence* 180: 35–37.

Hall, B. L. 2006. Social movement learning: Theorizing a Canadian tradition. In *Contexts of Adult Education*, ed. T. Fenwick, T. Nesbit, and B. Spencer. Toronto: Thompson Educational Publishing.

Hall, B. L. 2004. The kitchen, the shop floor and the street: Social movement learning for the 21st Century. In *Proceedings of the Joint International Conference of the Adult Education Research Conference and the Canadian Association for the Study of Adult Education*, ed. D. E. Clover, 13–18. Victoria: University of Victoria.

Hall, B. L. and D. E. Clover. 2005. Social movement learning. In *International encyclopedia of adult education*, ed. L. English, 584–89. London: Palgrave Macmillan.

Hill, R. J. 2006. Environmental adult education: Producing polychromatic spaces for a sustainable world. In *Global issues and adult education*, ed. S. Merriam, B. Courtney, and R. Cervero, 265–77. San Francisco: Jossey-Bass.

IPCC. 2007. *Summary for Policymakers of the Synthesis Report of the IPCC Fourth Assessment Report*. Geneva: IPCC Secretariat.

IPIECA. http://www.ipieca.org/activities/climate_change/downloads/workshops/26sept_06/Report.pdf.

Jickling, B. 2001. Editorial. *Canadian Journal of Environmental Education* 6: 5–7.

Keough, N., C. Emman, and L. Grandinetti. 1995. Tales from the Sari-Sari: In search of Bigfoot. *Convergence* 28 (4): 5–11.

Kumar, S. 2007. Ecology and economy. *Resurgence* 244: 3.

Mace, D. 2005. *A change of mind: Social movement learning and the cancellation of the Duke Point Gas Plant (Vancouver Island and Gabriola Island, 2005)*. [M.Ed. thesis, University of Victoria.]

Miller, K. http://www.kristinmillerquilts.com/Protest_Quilts/Positive_Energy_Quilts/Positive_Energy_Quilts.htm (accessed June 22, 2009).

Moore Lappé, F. 2009. Liberation theology: Beyond the disempowering message 'gospel' of 'no growth' to the ecological reality of 'endless possibility.' *Resurgence*, 252: 19–20.

Naniamo Citizens Organizing Committee. http://members.shaw.ca/NCOC/Frameset1/1/frameset1.html

Reimer, M. Forthcoming. Community psychology, the natural environment, and global climate change. In *Community psychology: In pursuit of liberation and well-being*. 2nd ed., ed. G. Nelson and I. Prilleltensky. New York: Palgrave.

Speth, J. G. 2008. *The bridge to the edge of the world: Capitalism, the environment, and crossing from crisis to sustainability*. New Haven: Yale University Press.

The Land Conservancy. http://www.conservancy.bc.ca.

Transformative Learning Centre. 1994. *Awakening sleepy knowledge: Transformative learning in environmental action campaigns*. Toronto: OISE/UT, TLC.

Wente, M. 2009. Good new, yep, good news. *Globe and Mail*, January 30, A15.

10 Transforming the Ecological Crisis
Challenges for Faith and Interfaith Education in Interesting Times

Toh Swee-Hin (S. H. Toh) and Virginia Floresca Cawagas

INTRODUCTION

Worldwide, there is now increasing acknowledgment of the serious problem of climate change by all sectors, from environmental movements and other civil society organizations to governments and intergovernmental agencies, including the United Nations. Unless urgent initiatives and efforts are implemented at individual, institutional, and societal levels to address climate change, our planet will suffer the impact of various facets of ecological destruction with dire consequences for many millions of human beings and countless other species. In these 'interesting' times, faiths, as much as other societal institutions and stakeholders, are challenged to show their relevance in facing the dangers of climate change and their capacities to seize the opportunities for becoming part of the solutions to this monumental crisis.

In this regard, a growing number of faith communities and institutions have joined the voices of concern and advocacy for action to resolve the urgent problem of climate change as well as the multiple dimensions of the ecological crisis. Thus, as early as 1992, the World Council of Churches (WCC) had formed a Working Group on Climate Change at the UN Rio Earth Summit. Steadily strengthening its advocacy, the WCC submitted a statement entitled "Climate justice for all" to the High-Level Ministerial Segment of the UN Climate Conference in Nairobi in 2006. Very recently, the world's largest and most representative interfaith organization, Religions for Peace, convened in Japan to prepare a statement urging G8 leaders to address the urgent problems of violent conflict and climate change at their July 2008 meeting in Hokkaido. Substantive progress has also been made by the Alliance of Religions and Conservation (ARC) in motivating faith institutions worldwide to implement the Seven Year Plans for Generational Action on climate change and on the environment by following guidelines drafted by the ARC-UNDP (United Nations Development Program) partnership.

These and many other similar exemplars of local, national, regional, and international advocacy by faith communities and interfaith organizations are hopeful signs that faiths and spirituality traditions are assertively facing the escalating problem of climate change. More broadly, while the urgency of addressing climate change is well recognized, faith leaders and networks have focused on the need for policies and individual practices to reflect the values and principles of ecological sustainability, including finding alternatives to overconsumerist lifestyles and unsustainable economic development.

This chapter therefore seeks to clarify this growing positive role of faiths and faith institutions throughout the world in educating and mobilizing members of their communities and through interfaith collaboration to address climate change and other facets of the ecological crisis that threaten the very survival of humanity. It will be shown how the teachings of diverse faiths and spirituality traditions reflect shared values and principles for inspiring their followers to live in harmony with the earth, and thereby contribute to the building of sustainable futures based on active nonviolence, justice, human rights, and intercultural respect. Exemplars will also be given of grassroots practices and initiatives showing how diverse faith communities are helping to overcome climate change and the ecological crisis through creative intrafaith and interfaith education.

FAITHS AND SUSTAINABILITY: VALUES, PRINCIPLES, AND TEACHINGS

Concomitant with the growth of modern awareness of environmental destruction and the dire consequences for human and planetary wellbeing, theologians, leaders, and practitioners of diverse faiths worldwide have increasingly searched for and revealed the ecological wisdom deep within their traditions (Tucker and Grim 1994; Barnhill and Gottlieb 2001; Barnes 1994; Ruether 1992; Mische 2006; Bingham 2009).

In the words of Denis Edwards (2006, 2):

> Religious faith has an important contribution to make to the ecological movement. It can give meaning and motivation, build an ecological ethos, and contribute to the foundations of an ecological ethics. For many people around the globe religious faith continues to provide fundamental meaning. For such people, ecological commitment can receive its deepest grounding only at a religious level.

In a similar vein, though with a helpful caveat, Tucker and Grim (1994) note:

> In addition, the emerging dialogue on religion and ecology also acknowledges that in seeking long-term environmental sustainability, there is clearly a disjunction between contemporary problems regarding the environment and traditional religions as resources. The religious traditions are not equipped to supply specific guidance in dealing

with complex issues such as climate change, desertification, or deforestation. At the same time one recognizes that certain orientations and
values from the world's religions may not only be useful but even indispensable for a more comprehensive cosmological orientation and environmental ethics. (1994, 2–3)

As a growing number of theologians and faith leaders and elders have artic-
ulated, there is much common ground in values, principles, and teachings
across faiths and spirituality traditions that promote living in harmony with
the earth despite some distinct and unique sources of faith and spiritual
inspiration. One key shared understanding rests on the concept of interdependence among all parts of the world or universe, including human beings,
other sentient beings, and the natural environment. Thus for indigenous
peoples, the wisdom of their elders has through rituals, ceremonies, daily
practices, and traditional knowledge contributed to sustainable relationships
between peoples and the land and all other parts of "mother earth" (Knudtson and Suzuki 1992; Grim 2001). As part of a holistic web of life, human
beings have a duty to respect, nurture, and sustain not just ourselves but also
other beings and nature, for in an interconnected world, all would suffer
the consequences of any action that diminishes the web's wellbeing. Among
recent ecological statements, the *Earth Charter* (2000) has helped to enhance
recognition of the ecological wisdom of indigenous spirituality.

In the Hindu tradition, many rituals likewise symbolize a deep reverence for and interconnectedness with all parts of nature. The Hindu worldview that all reality and everything in the universe, including humans and
nature, are all sacred as a part of "God" and sustained by divine forces is
a clear reminder to Hindus to treat all beings and all parts of nature and
the environment with deep respect and compassion (Prime 2006; Nelson
1998; Chapple and Tucker 2000). In a parallel way, Jainism's core belief
that every living being has a divine soul with the innate potential of being
liberated from karma through mental and physical purifications calls on
believers to practice *ahimsa* and compassion to all beings (Chapple 2002;
Kumar 2003). For Daoism, the central concept of the 'Dao' reflects the
fundamental unity and interdependence of all beings, nature, and the entire
universe (Girardot et al. 2001; Chen Xia 2006). To be in accord with the
Dao, human beings need to live in harmony with nature, although nature is
not an entity (e.g., wilderness) separate from culture and hence also needs
to be cultivated.

As revealed in recent anthologies based on sutras and other texts, Buddhist teachings encompass key principles, concepts, and values in accord with
ecological thinking and action (Kaza and Kraft 2000; Tucker and William
1998). Indeed, one of the most basic concepts in Buddhism, interdependent
co-arising, whereby all things and phenomena are interdependent, reminds
humanity that we are integrally interrelated with all beings and all parts of
the universe (Thich Nhat Hanh 1998). This principle of interdependence,

together with the Buddhist values of loving-kindness and compassion, further guides practitioners to relate to all other beings and parts of the natural environment in a peaceful and caring way (Swearer 1998).

Another major principle that supports an ecological ethic rests on God's divine act of creation of all beings and the universe. Thus in Judaism, the very first chapter of Genesis reveals that all parts of God's creation have their own intrinsic value, and emphasizes the holy interconnectedness of humanity and nature (Bernstein 2005; Tirosh-Samuelson 2003). This principle calls on Jews to treat all other beings with compassion, rather than acquisitiveness or aggression. Among environmentally concerned Christians, the theme of 'creation spirituality' has also been central. Thomas Berry's influential *The Dream of the Earth* (1988), for example, emphasizes an understanding of creation whereby all elements including human beings and nature are intimately interrelated in the wider earth community. In proclaiming St. Francis of Assisi the "Heavenly Patron of Ecologists" on Easter Sunday in 1980, Pope John Paul II affirmed St. Francis' inspirational vision of deep love and respect for nature, and living in sustainable relationships with all parts of creation (Sorrell 1988). Another Catholic theologian, Denis Edwards (2006), argues that not just humans, but all creatures reflect the image of God, and hence theocentrically, human beings are called to love and respect the integrity of each creature, and to care for God's creation. For believers in Sikhism, nature too is conceived as infused and sustained by the divine presence of the One Creator, thereby deserving of the gratitude and reverence of all Sikhs (Singh 2008). In the Bahá'í faith, there is similarly a core belief in seeing nature as God's Will and creation and in the interdependence of all beings (Dahl 1995).

Furthermore, the call for environmental care among Jews and Christians rests on an interpretation of the Biblical concept of 'dominion,' which is not exploitative but rather means responsible stewardship of all parts of God's creation. For Jews, this is affirmed by the teachings of *bal tashhit* (do not destroy wantonly nor waste resources), *za'ar baalei hayyim* (pain of living things), and the sabbath, a reminder that creation is beyond economic considerations (Ehrenfeld and Bentley 1985, Sears). The emergence of what is now referred to as 'green theology' in the Christian community has provided critical interpretations and rereadings of the scriptures that support the concept of 'ecological conversion,' whereby Christian faith shifts from earlier conceptions of 'dominion' to 'stewardship' (Berry 2006) and is translated into environmental awareness and action. Among evangelical Christians, "environmental stewardship is therefore a matter of both Christian obedience and Christian piety" (Bullmore 1998, 141). The Orthodox Church believes that "greed to satisfy material desires leads to the impoverishment of the human soul and the environment" (Message of the Primates of the Orthodox Churches 2008).

For the growing number of Muslims worldwide promoting environmental care and sustainable futures, the Prophet Muhammad (pbuh) is an inspirational and pioneering role model for conservation and sustainable resource management (De Chatel 2003). An ecological ethic within Islamic teachings is based on four central principles and beliefs (Khalid 2002). The first, *Tawhid* or the Unity Principle, proclaims there is only one God who has created everything. Secondly, the creation principle, *Fitra*, upholds the unity of all creation of which humanity is one part of the natural inter-connected order created to serve the will of God. According to the third principle, balance or *Mizan*, all parts of creation are endowed with a role and purpose and are related in a state of dynamic balance. Fourth, and most crucially from the perspective of environmental sustainability, there is Khalifa or the responsibility principle which upholds the role of trustee-ship (*amanah*) for the wellbeing of humanity and all nature. Citing exten-sively from the Qu'ran, Ozdemir (2003) likewise demonstrates a strong case for environmental ethics in Islam, including the key principles that all nature is Muslim and that human beings as vice-regents of God on earth are accountable for their actions toward the environment.

These shared principles of diverse faiths that underpin an ecological ethic are also complemented by other beliefs, values, and practices supporting sustainable relationships. These include, for example, simplicity of living and the importance of searching for inner peace in Hinduism and Jain-ism. As Gandhi emphasized, a nonviolent world is promoted by reduced consumption, self-reliance, sustainable practices, voluntary simplicity, and spiritual growth. Early Daoists have implemented environmentally respon-sible rules, rituals, and policies, including simplicity and self-reliance in living, restraining desires and power-seeking, and meditation for health and wellbeing.

Traditionally, Buddhism has stressed the need for individual cultivation and practice of the Eightfold Path and other Buddhist teachings as this is indispensable from the viewpoint of karmic self-responsibility. More recently, as reflected in the views of "engaged Buddhists" (Thich Nhat Hanh, 1998; Sivaraksa 2005; Jones 2003), a Buddhist ecological ethic nec-essarily also requires Buddhists to act for societal transformation, akin to the approach of Christians concerned with structural and systemic changes for peace and justice (e.g., liberation theology and other movements focus-ing on the social teachings of the Christian faith).

Likewise, the Sikh Gurus integrated the principle of living in harmony with the environment by building temples (gurdwaras) with community water tanks and flower and fruit gardens and practicing vegetarianism (Sikhism and the Environment). The institutionalization of shariah law in the Muslim world has also included environmental conservation and protec-tion elements such as moderate consumption, charity to animals, responsi-ble utilization of scarce resources, state-established conservation areas, and conservation-oriented charitable endowments (Khalid 2002; Bagader et al. 2006). While lauding material development, Bahá'u'lláh cautioned against excessive materialism, which parallels the environmental ethic of moderate consumption. Inspired by the social teachings of their faith, Christians are called to live simply so that others can simply live. Hence ecological conversion necessarily implies a strong sense of ecological justice.

FACING CLIMATE CHANGE, CREATING SUSTAINABLE FUTURES: THE HOPES AND CHALLENGES OF ECOLOGICAL CONVERSION

In his helpful overview and in-depth reporting on the positive contributions of religions and faiths toward promoting sustainable development, Gardner (2006) referred to five assets that faith communities and institutions bring to these efforts: meaning, moral capital, numbers of adherents, land and other physical assets, and social capital. In his view, providing people with a sense of "meaning and purpose" is one of the most powerful assets wielded by religion. While Gardner appropriately cites sacred storytelling, symbols, and rituals as sources of such ecological meaning, our concise exploration of values, principles and beliefs across diverse faiths in the preceding pages suggests a more holistic way to conceptualize this asset. Faiths will need to include in their social education and learning processes a systematic theological awareness of the ecological values and principles already rooted deep within their cosmologies. As shown earlier, this may need a critical rereading or reinterpretation of scriptures and other sacred texts.

A second key asset is "moral capital," whereby faith leaders are able to exert their moral authority and influence into the wider social and political realms of their societies and the wider world. For Gardner, this asset's potential remains undertapped, primarily since faith leadership at the local, grassroots levels may still not be convinced of the need for particular advocacy or transformation.

Third, religions have considerable numbers of adherents. Possibly over 80 percent of the world's population belong to over 10,000 faiths and religious institutions spread over diverse regions. If and when adequately educated and inspired to act, these numbers, even if only a portion, can help move the sustainability agenda forward.

The fourth asset Gardner lists is the control by religions of substantial physical and financial resources, including land, buildings, and wealth that in many cases, especially in North contexts, are invested. These investments constitute potential political leverage (for example, as seen during the anti-apartheid dis-investment campaign).

Finally, there is the asset of social capital, "the bonds of trust, communication, cooperation and information dissemination that create strong communities" (Gardner 2006, 52) that are essential for successful societal transformation.

To these five assets, however, a further asset-cum-institutional capacity is necessary in our view, namely the requirement of a transformative social learning paradigm for all faith communities and institutions. One key component of this paradigm critically educates faith followers, including leaders, on the root causes of the ecological crisis, including why and how climate change has become such a threat to human and planetary survival, together with the range of possible solutions to specific components of the environmental problem. Faith actions for sustainability need to be guided by the best available scientific evidence, such as the recent UN panel reports and related studies on climate change.

Equally essential, however, is another component of social learning that has been integral in our work on educating for a culture of peace (Toh 2000; Cawagas 2006). This is the requirement of key pedagogical principles to enable faith values and sacred teachings and environmental science knowledge to be effectively translated into personal and social action. Concisely, these principles include:

- *holism,* whereby ecological issues and problems are understood in their interrelationships with other themes of conflict and violence (e.g., local/global injustices, human rights violations, wars), and nonformal, informal, and formal modes of learning are simultaneously pursued;

- *dialog,* which requires participatory, horizontal, and creative modes of learning rather than traditional top–down, one-way 'banking' models;

- *values-based education* that acknowledges the presence of values in all learning contexts (something already transparent in faith learning environments) that are sources for the intrinsic reinforcement essential in social learning; and

- *critical empowerment,* whereby learners are empowered to translate critical awareness and consciousness into principled and sustained action for personal and social transformation.

For a number of faith communities and institutions, some of these critical pedagogical principles are being applied in their initiatives to build sustainable futures. Thus Christian environmentalists like Gardner (2006) and McDonagh (1986) have exemplified praxis in the spirit inspired by an ecological conversion grounded in ecclesial hope and solidarity, while engaged Buddhists like Sivaraksa (2005), Thich Nhat Hanh (1998), and Jones (2003) cultivate mindfulness and compassion simultaneously at individual and societal levels. However, due to institutional legacies or organizational orientations, other institutions may still require substantive changes in policies and practices before their learning frameworks attain a transformative quality. For example, faith leaders may prefer established modes of change based on hierarchical directives with little space for participatory decision-making and social learning. Empowerment may also be confined to individual improvement while social or structural transformation is avoided in order to distance faith from politics.

COMMUNITY-BASED EXEMPLARS

In recent years, concomitant with the public concern, debate, and governmental and intergovernmental efforts for national/international action, diverse faith and interfaith organizations and communities have increasingly identified their programs under the umbrella of climate change. However, prior to our reflection on several community-based exemplars, it is necessary to first acknowledge the dedicated and courageous ecological praxis of many faith and interfaith communities, NGOs, and CSOs (civil society organizations) that has been ongoing in specific areas of environmental social learning and action, long before 'climate change' became a major referent or focal point. Not to do so would be to contradict the values of gratitude and thanksgiving deeply endorsed by all faiths, for it is the multiple seeds of past action that bear fruit today to inspire current campaigns. Also, continuing facets of ecological destruction that may not be directly linked to climate change continue to inflict great suffering on humanity globally and hence deserve continuing education and action.

For example, in virtually all continents, indigenous peoples have been engaged in environmental campaigns to challenge the negative impact of mining operations on their ancestral lands. Often run by transnational corporations, these mines have caused much environmental damage to land, water, and air resources on which indigenous peoples' livelihoods and cultural survival depend, leading on many occasions to territorial displacements and other human rights violations (Alejo 2000; Mining Watch). In late March 2009, eighty-five delegates from thirty-seven countries gathered in the Philippines at the International Conference on Indigenous Peoples and Extractive Industries to appeal to the UN, multilateral agencies, and governments to address the problem (CBCP News). Moreover, the mining industry is a significant emitter of greenhouse gases and hence a direct contributor to climate change.

In the area of food security, the determined actions by poor farmers in India, together with environmental advocates such as Vandana Shiva (2005), show how traditional Hindu and other indigenous beliefs and practices in people-land-food relationships can help build an "Earth Democracy" to overcome the ecologically destructive effects of globalized, export-oriented, monoculture, agribusiness-dominated, and hi-tech food production systems. Instead, indigenous seed banks, organic farming, community-controlled irrigation, farming cooperatives, microcredit schemes, and land reforms provide grassroots alternatives that reduce inequalities, meet basic needs, and promote ecologically sustainable agriculture.

In Africa, the Africa Muslim Environment Network implemented a project to help Muslim fishermen along the Kenyan coast to give up destructive fishing techniques and return to traditional fish trapping and to market the ecologically harvested fish to hotels and resorts. This and another similar project for Tanzanian fishermen "are experimental models for adopting Muslim education and environmental ethics as a way of innovative conservation and slowing the depletion of fish resources" (Sam Aola Ooko 2008). Clearly, as climate change impacts negatively on food security, such faith-based and cultural ecological alternatives in agriculture and fisheries must be urgently pursued.

Turning to another indispensable and climate change–affected resource for life, namely water, various faith communities have been drawing on faith wisdom and indigenous knowledge to help conserve water resources and promote more sustainable water systems. As Shiva (2002) has cogently analyzed, forces of excessive privatization and profit-centered exploitation of water resources are aggravating water shortages, impacting especially on poor peoples. She shows how Hindu and indigenous traditions and knowledge have been helping to revitalize water sources based on community democracy and equitable access. Given the very serious state of the global water crisis (Barlow 2007), and since water holds a central sacred place in diverse faiths, the potential for faith communities to be active in instituting and implementing clean water laws needs to be tapped (Fisher-Ogden and Saxer).

Another major and long-standing focal point for environmental action, which continues to be a vital tool in addressing climate change, is in protecting the remaining rapidly diminishing forests and in reforestation. The courageous struggles of the Chipko tribals in India to save their forests (Webber 1989) and the success of the Kayapo peoples to gain sovereignty over at least a portion of their ancestral domains in the Amazon (Garrigues 2008) are two of numerous exemplars of indigenous spirituality translated into ecological praxis throughout the world.

No less inspiring is the story of basic Christian communities in the southern Philippines who engaged in courageous nonviolent action (pickets and lobbying government) against environmentally destructive logging (Gaspar 1990).

In all these exemplars of faith-based ecological praxis, the pivotal role of critical social learning is indisputable, awakening marginalized peoples to the root causes of unsustainable and exploitative unjust development, and catalyzing committed social action for ecojustice.

GREENING OF FAITH SPACES

In many countries, faith institutions and communities are joining ecumenical and interfaith movements and networks that seek to address climate change by greening their spaces and infrastructure. In the United Kingdom, for example, the Living Churchyards and Cemetery Project facilitates local communities to be involved in redesigning their churchyards and cemeteries for the benefit of wildlife. In Canada, the interfaith network, Faith and the Common Good, has initiated a Greening Sacred Spaces program "to assist faith communities in their efforts to renew the Sacred Balance by making their communities more harmonious with the natural world" (Faith and the Common Good).

Each participating community commits itself to implementing greening practices (such as recycling, nondisposable dishes/cutlery, ecofriendly office and building supplies, Fair Trade products); organizing outreach activities (such as conferences, ecologically-oriented prayer services, eco-awareness and action); and building retrofits (such as energy efficient lighting, insulation, and appliances; renewable energy; energy audit; low-flow toilets). Apart from the cost savings, the faith communities often were able, through education and role modeling, to influence individual members to 'green' their own homes and other spaces. A similar and even more systematic program has been set up by the Church of South India in Kerala State based on a long-range comprehensive plan (Alliance of Religions and Conservation).

In the United States, one of the most successful and rapidly expanding movements linking faith and ecology is the Interfaith Power and Light (IPL) program of the Regeneration Project founded by the Rev. Canon Sally Bingham, environmental minister for the California Episcopal Diocese. Involving now more than 4000 congregations and over 500,000 people in twenty-eight states, the IPL's work includes educating congregations and helping them buy energy-efficient lights and appliances, energy audits, using more fuel-efficient vehicles, supporting renewable energy development, and advocating for sensible energy and global warming policy.

Another influential US initiative is the Interfaith Climate and Energy Campaign, a coalition of religious American leaders, institutions, and individuals working to educate their members about the causes and effects of climate change and to advocate for effective governmental policies (Earth Ministry 2003).

There has also been increasing interest among Muslim leaders worldwide to integrate climate change and other environmental issues into their religious duties, such as an imam in a London mosque whose Jumah sermons encourage responsible energy and water use (Burgess 2007) and a Malaysian imam teaching about and coordinating a turtle conservation project.

In Australia, following on the Climate Institute's compilation of statements on climate change by diverse faith institutions, an Australian Religious Response to Climate Change (ARRCC) network was established in late 2008. ARRCC encourages faith communities to educate members to take action at institutional and home levels to switch to an eco–power provider, reduce energy consumption, and join campaigns for stronger governmental policies to reduce CO_2 emissions.

Among various Australian faith institutions, there is a growth of faith-inspired projects to address climate change issues, such as Brisbane's Catholic Justice and Peace Commission's Cool Communities project encouraging households to reduce electricity and fossil fuel consumption (Arndt 2005). Under the leadership of Imam Afroz Ali, the Sydney-based Al-Ghazzali Centre is developing a program with the local city council to establish, protect, and maintain the natural ecosystems and wildlife corridor to regenerate an environmental disaster zone.

The Griffith University's Multi-Faith Centre, established in 2002 to promote interfaith dialog for a culture of peace, has consistently integrated into its social learning and action programs themes related to overcoming the ecological crisis and building sustainable futures. Specific activities include a year-long workshop with Catholic school teachers to introduce an interfaith perspective in a new curriculum for a culture of peace including building sustainable futures; commemorating World Environment Day and Earth Hour; an interfaith dialog training program for forty local faith and interfaith leaders; and international symposia and summits on such themes as Cultivating Wisdom for a Culture of Peace; Women, Faith and a Culture of Peace; and One Humanity, Many Faiths.

Globally, the work of the Alliance of Religions and Conservation (ARC), a "secular body that helps the major religions of the world to develop their own environmental programmes, based on their own core teachings, beliefs and practices . . . [and] link with key environmental organisations—creating powerful alliances between faith communities and conservation groups," has yielded many positive and inspiring outcomes (ARC). For example, the Sacred Lands project, initiated in the United Kingdom "involved Buddhist, Christian, Hindu, Jewish, Muslim and secular communities in creating and reviving inner-city and community gardens; conserving and celebrating holy wells; rediscovering and renewing pilgrimage trails; protecting trees and woodlands; regenerating community meeting places and their eco-systems; and celebrating sacred places with works of art and poems." Recognizing the vision that faith-owned forests should be run on ecologically sustainable and socially just lines, ARC convened a Forests and Faiths meeting in Visby, Sweden (August 2007), which agreed to create a Religious Forestry Standard to be launched in 2013.

Most significantly, ARC's partnership with UNDP to involve major traditions in the world's faiths to design and launch the Seven and Eight Year Plans for Generational Change on climate change and the environment emphasizes the need for longer-term transformation. Under these plans, participating faith communities will systematically implement practices and protocols for various areas including energy use, sacred sites, education, building programs, and celebrations. Given the widely acknowledged environmental crisis facing China, it is significant that the first such plan has been created by its Daoist community. Specific proposed activities in the China Daoist Ecology Protection Eight Year Plan include more environmentally friendly pilgrimages to Daoist sites, protecting Daoist sacred mountains, reviving Daoist health-promotion practices, Daoist ecological education, youth camps, promoting simple and energy-saving lifestyles, and ecological integration in rituals and festivals.

The steady growth in interest in these Seven or Eight Year Generational Plans with ARC's encouragement across faiths and regions (such as the Seven Year Plan for Muslim Action on the Environment for Islamic states and Muslim communities; Buddhist monks in Cambodia and Mongolia; and Bahá'i community's International Environment Forum) is a hopeful sign for coordinated global faith-based action on climate change and building sustainable futures.

INTERNATIONAL AND GLOBAL ADVOCACY

In facing the climate crisis, clearly there is a need for concerted international and global policy shifts by governments and intergovernmental agencies. Together with 'secular' environmental NGOs and networks, many faith institutions, and increasingly interfaith movements and organizations, have been seizing opportunities to be heard at the tables of previous and forthcoming international consultations and decision-making conferences on climate change.

The World Council of Churches (WCC), for example, through its Working Group, has continued since 1990 to have a sustained presence at all UN negotiations on climate change. In the post-Kyoto era, the WCC has been calling for comprehensive policies for adaptation programs in countries severely affected by climate change. One focus for heightened attention is the impact of climate change on water as reflected in the formation of an Ecumenical Water Network (EWN) on the rights to water and community-based initiatives.

As mentioned in the introduction of this chapter, the global interfaith organization Religions for Peace also presented, through the Japanese Prime Minister, a statement urging G8 leaders to address the urgent problems of violent conflict and climate change at their July 2008 meeting in Hokkaido.

This international and global advocacy by faiths and interfaith organizations, together with public campaigns at local and national levels, represents an essential dimension of the role that faiths can and need to play in influencing and challenging governments and intergovernmental agencies to take bold steps in addressing climate change and the ecological crisis.

FORMAL EDUCATION EXEMPLARS

The role of formal education institutions in promoting faith- and interfaith-based curricula and learning to address climate change and the wider ecological crisis is undoubtedly crucial in complementing nonformal and informal community-based programs. Children and youth who gain a deeper knowledge of environmental destruction and are empowered to take personal and social action to build sustainable futures will grow up to be better prepared and motivated adult citizens to join the global cause of saving the earth and humanity. Higher education graduates in diverse professions and, from the perspective of this chapter, faith leadership fields similarly need sufficient grounding in ecological understanding, skills, and commitment.

Across the world, national educational systems have been integrating education for sustainable development or sustainability into formal curricula and extracurricular areas. In recent years, the UN Decade of Education for Sustainable Development (2005–14) is helping to generate momentum in this field of educational innovation, with lead agencies such as UNESCO playing a service and mobilizing role in curriculum development, teacher education, and policy formulation (Fien 2006). Students in faith-based schools which are part of the national systems are required to follow standard curricula guidelines, hence will be as much exposed as students in non-faith-based state schools to climate change and related issues.

However, in contrast to community-based exemplars, the literature available in the public domain on role models of schools of diverse faiths active in faith and ecology and climate change education is smaller in volume. This may well be because schools and teachers do not necessarily publish reports on their daily teaching and learning activities for WWW or print media distribution. Nevertheless, a number of interesting and inspirational examples have been documented across faiths and regions.

In Brazil, for example, a private Catholic sisters' school has been able to weave ecological and ecojustice perspectives throughout the formal and extra-curriculum, drawing in part on the *Earth Charter*, including linking students with nearby slum communities to assist in water quality testing, waste reduction, and recycling, researching alternative energies, and summer volunteer work teaching about recycling and composting in poor neighborhoods (Gardner 2006, 73–74).

In the United States, primary curricula on climate change have been developed for Christian primary schools (Catholic Schools and Climate Change). In Boston, the Shalom Center has published a curriculum and ceremony for bar/bat mitzvah and confirmation-age youth and families "on how the younger and older generations can work together to heal the earth from the dangers of global climate crisis" (The Shalom Center).

In Indonesia, after the peace accord that ended the armed conflict in Aceh, new peasantren, or traditional Achenese Muslim schools, are playing a leading role in cooperation with surrounding ecocommunities to replant forest lands destroyed by years of civil war (ARC). There is a growing number of peasantrens where Islamic values and teachings related to ecology and environmental sustainability are being taught (Gelling 2008). Recently, the first state-funded Hindu school in the United Kingdom and Europe

opened, based on an ecological curriculum and ethos, integrating meditation, a vegetable garden, vegetarianism, solar panels, rainwater recycling,
and teaching in the school grounds (ARC).

In the Philippines, the Assumption College of Antipolo integrates creation
spirituality in its curriculum and promotes ecofriendly practices, including
solid waste management, energy efficiency, a PACEM Eco-Park featuring an
ecology center, mini–rain forest, butterfly garden, aviary and wildlife sanc-
tuary, herbal garden and organic farm, and "Adopt A Piece of the Planet"
project in which each class takes care of a particular spot on campus.

In Australia, there are Catholic schools engaged in watertank culture,
permaculture, and energy efficiency projects, while a Catholic school
curriculum framework deserves to be lauded for its holistic and critical
approach. Designed under the auspices of Catholic Earthcare Australia, *On
Holy Ground: An Ecological Vision for Education in NSW* (New South
Wales) not only integrates knowledge about the ecological crisis with the
Church's call for ecological conversion based on principles of sustainability
and justice, but also empowers students for personal and social action.

In sum, these exemplars suggest that faith-based schools have considerable potential in preparing children and youth to be active local and
global citizens in facing climate change and other dimensions of the ecological crisis, and in building a more sustainable world. As the Alliance
on Religions and Conservation notes, the recent report *Leading Sustainable Schools* by WWF-UK for the National College for School Leadership confirms that faith schools and colleges have accomplished much and
are most active in developing sustainable activities and practices due to
leadership commitment and passion inspired by their faith's spirituality
(Jackson 2008).

Turning to the important area of formation or training of faith leaders, whether in seminaries, theological colleges, monasteries, madrasahs,
or synagogues, climate change and the wider ecological crisis is undoubtedly a vital component in their formal education programs. Clergy, imams,
rabbis, monks, nuns, and other faith leaders need formal grounding in climate change and environmental care in the context of their faiths' teachings and wisdoms, as well as critical pedagogy skills that will allow them
to empower their future congregations and parishioners.

For faiths whose leadership and institutions are already committed to
ecological conversion and sustainable futures, this need for well educated
future faith leaders in the range of religious formation bodies would be
acknowledged and implemented. Increasingly, formal theological curricula
are including courses on climate change and the ecological crisis. One interesting US-based example is the Green Seminary initiative, which builds on
the work of Theological Education to Meet the Environmental Challenges
to "build a broad, diverse, active coalition of faculty, administration, and
students committed to creating sustainable seminary communities that
will model care for creation and that will equip pastors and other religious

leaders with the tools needed to lead their congregations and communities in responding faithfully to the ecological challenge" (Donofrio 2007).

Recently in Thailand, twenty-five Thai Buddhist monks and nuns undertook a nine-month training program sponsored by the UK-based Artists Project Earth group so that they may be able to empower youth and villagers to be ecologically aware and to make concrete plans for "redesigning monasteries to be spiritually-based eco-learning centres transitioning to renewable energy" (Artists Project Earth). As earlier mentioned, the China Daoist Eight Year Ecology Protection Plan includes ecological training for Daoist priests.

As the work in linking faith and ecology proceeds, faith leaders are increasingly called on to be community educators for empowering their faith members to challenge climate change and engage in transformation.

BE HOPEFUL, BE MINDFUL

This overview and exploration of the role of faiths in addressing climate change and the wider ecological crisis has, in our view, revealed many possibilities and signs of hope. Across the diversity of faiths and spirituality traditions, there is a growing awakening to the embeddedness within the teachings and sacred texts of deep values, principles, and wisdom consonant with an ecological ethic. This recognition has been translated into a broad range of creative, dynamic, and multifaceted community-based educational projects and programs in many countries and regions.

Through the social learning processes of these activities, members of faith communities have become engaged in ecological praxis at local, societal, national, international, and global levels to face the climate crisis and build alternative sustainable futures. Even where particular faith communities or traditions may not necessarily participate in the wider project of interfaith dialog, due to existing more exclusivist orientations on their 'truths,' nevertheless it has been possible for such communities or denominations to cooperate with other faiths in meeting the challenges of climate change. The complementary role of formal education programs and institutions is also visible and encouraging, especially in schools and focused higher education programs. The extent to which faith-formation institutions are becoming more ecologically sensitive needs further investigation.

On the issue of the relative importance or balance between formal, nonformal, and informal social learning, we are inclined to place priority in three sectors: community faith-based contexts, formal faith leaders' formation/training or in-service institutions and programs, and tertiary learning institutions as well as senior high schools. Adult citizens, including leaders, who profess a faith identity can be moved to reconsider their personal and social practices when they consistently hear messages and wisdom from their faith leaders and institutions of the dangers of not addressing climate

change, witness the leaders modelling 'green' conduct, and most importantly, if the social learning process is critical and empowering, engage in
personal and societal transformation within and beyond their faith communities. However, this will in turn need a pool of faith leaders appropriately
educated, skilled (reskilled), and spiritually motivated in social learning
and transformation to overcome climate change. At the upper levels of secondary schooling, and continuing into tertiary institutions, ecologically
awakened learners will be able to join the wider society as young adults
inspired by the ecological wisdom of their faiths to serve as committed
advocates for sustainable futures.

Another feature of these 'interesting' times is of course the ubiquitous
exponentially expanding dimension of information technology (IT). Whether
in community or formal social learning processes, the positive and influential role of IT and other media tools in spreading knowledge about climate
change, sharing ideas and exemplary practices for building sustainable relationships, and strengthening networks for local and global advocacy is now
widely acknowledged. This does not mean, though, the end of face-to-face
social learning processes. As the experiences of peace education and other
transformative educational fields (for instance, social justice, human rights,
intercultural understanding, disarmament, education for sustainable development, and gender equity), and indeed interfaith dialog and education, have
shown, the opportunities for human beings to be together as part of the 'one
world' family, directly sharing despairs but also the joys of small successes,
and hopes and dreams, are invaluable catalysts for life-changing moments.

However, while affirming these hopeful signs of progress, it is important
to remain mindful of the ongoing obstacles and barriers facing faiths in fully
realizing their potential role in resolving this urgent global crisis. Thus, for
every example that has been listed in this chapter describing fruitful and successful work of a faith community or institution to address climate change,
in reality there will be many more cases where little is being done to bring
the problem and issues of climate change and the ecological crisis into the
vision, mission, and life of the faith community. While senior faith leaders
can and have issued strong inspirational messages for ecological conversion,
it is always a slow and patient educational process for the vision to be fully
embraced throughout the grassroots of all faith communities. From a social
learning perspective, it is of course vital for faith leaders to be inspirational
role models for ecological praxis. Hopefully, the pace will quicken as newer
faith leaders emerge from more ecologically-friendly basic schooling and then
faith training institutions. Likewise, young adults in the days ahead will be
leaving their schools more attuned to ecological praxis, and hence able to
'move' faith leaders to be more responsive to the crisis.

A final crucial barrier to be faced and overcome lies in a more foundational or paradigmatic area, namely, how the climate and wider ecological
crisis is understood in terms of root causes and hence how really sustainable and transformative will the proposed 'solutions' be. In looking at the

current advocacy on climate change, clearly while it is vital and urgent to reduce carbon emissions significantly, is this 'solution' still framed within the dominant paradigm of economic 'progress' and 'development' led by the corporate, free-market profit-centred and globalization-from-above system, or will it be viewed as a necessary step of a distinctly alternative paradigm of how humanity needs to live in holistic and interdependent relationship with the earth and all other beings and parts of the universe? Already, in the midst of the current global financial crisis and recession, the orthodox solution of unlimited consumption including trillions of dollars of spending 'stimulus' is being promoted by almost all governments and economic elites, without much thought to the destructive ecological consequences. Despite official acknowledgments of the severity of the crisis, many governments have been reluctant to enact policies for the deep CO_2 emissions targets required, on the grounds of protecting jobs and economic growth.

The UN panels as well as the indisputably wide and solid consensus among environmental scientists throughout the world have revealed increasingly dire and cataclysmic scenarios for climate change. In this regard, faiths embody a rich store of deep values, wisdom, and spiritual strength that can inspire or intrinsically motivate humanity to critically rethink dominant and unsustainable notions of 'progress' that lie at the very roots of the climate change and ecological crisis. This is where faiths, if they are to be authentic witnesses to the core values of love, compassion, loving kindness, justice, sustainability, and solidarity with all those in suffering worldwide, will need to be assertive when engaging in dialog with governments and other powerful economic stakeholders. To avoid being inadvertently co-opted into an essentially technocratic paradigm of addressing climate change, faiths will clearly need to identify that structural violence of local/global injustices and the dominant paradigm of overconsumerism, excessive materialism, and the commodification of all life are at the deep roots of the climate change problem and wider ecological crisis. Hence the apt concept of 'climate justice' in response to 'climate change.' Faiths can then humbly offer the values, principles, teachings, and wisdom as guides for reflection and transformative action, by individuals, communities, nations, and global institutions, in pursuance of an alternative future of wellbeing for all generations of beings and all creation or the universe, where there is spiritual growth and cultivation in sacred balance and harmony with moderate lifestyles and our planet or mother earth.

To close we offer two possible social learning scenarios of faith communities responding to the climate and ecological crisis.

Scenario 1

A small island in the Pacific has been suffering the effects of climate change, leading to loss of low-lying coastal areas, flooding, drought, coral destruction, weather extremes, salination of fresh water, and fisheries and agricultural failures. Despite the efforts of the international community and a wide range of civil society organizations, including faith and inter-faith communities, no significant resolution of the crisis has resulted due to the unwillingness of major actors (governments, corporations, global north citizens) to bear the 'costs' of decisive cutbacks in CO_2 emissions. The future looks very grim and the people of the island know that soon they will have to abandon their home. With surrounding island nations in a similar plight, they turn their eyes to their nearest north neighbor, Australia, and prepare to go there as environmental refugees. In Australia, there is growing debate in government and the public about the situation. Some politicians and citizens are reluctant to accept these climate change refugees, arguing it would set a precedent as more islands meet the same fate. They mount a campaign based on fear and self-interest.

However, among the faith communities, many faith leaders and mem-bers who have been active in the campaigns of their faiths and interfaith movements to address climate change, begin an active grassroots social learning and educational project throughout Australia. Using the social learning contexts provided by faith sermons, media programs, workshops, petitions, rallies, prayer vigils, and other active nonviolent strategies, they call on fellow faith members and other Australians to live up to the Austra-lian ideals of 'fair go' and compassionate acceptance of human beings forced to flee from their homelands to the safety of Australia (as occurred during earlier periods of refugee settlement). They also note that Australia has been one of the top contributors to climate change. In faith-based schools and universities that have been integrating climate change in educating for active citizenship, many teach-ins and rallies are also organized to mobilize public support for the climate change refugees. As the islanders begin to approach Australia's coastline, the Australian Parliament sits to debate a bill rejecting their entry as refugees. Will the educational, social learning, and political efforts of the faith communities, who feel that Australia has a spiritual and ethical responsibility to welcome and assist strangers in such great suffering, succeed in the defeat of the bill?

Scenario 2

The climate change situation worldwide has slightly abated as a result of a willingness of all nations and their citizens, especially those in indus-trialized or industrializing economies, to agree to a UN Climate Change Framework at the Copenhagen Convention in 2009. This is in part due to the determined efforts of both secular and faith- and interfaith-based institutions and movements to educate all sectors of their societies and the world community on the grave dangers to human and planetary survival posed by climate change and the wider ecological crisis. In many countries, sermons and religious teachings in churches, mosques, synagogues, tem-ples, and other places of worship regularly include ecological themes and responsibilities. Interfaith centers play a leadership role in bringing faith leaders and communities together for conferences and workshops on how faiths share a common ground in ecological wisdom, and explore opportunities for joint social learning and action to persuade governments to enact strong targets on reducing carbon emissions.

Diverse faiths have also enrolled in the ARC's Seven or Eight Year Generational Plans on Climate Change and the Natural Environment. By 2013, several detailed reports on these plans have demonstrated that from senior to grassroots levels, faith institutions and communities are not just talking about the problem, but their spaces, assets, educational programs, and the conduct of individual members have been transformed to reflect ecological, social, and spiritual wellbeing. Places of worship in effect provide the 'green' models for the observational learning of their members, who are then motivated by their faith's ecological wisdom to change lifestyles and home spaces to reduce their carbon footprint. Partnerships with ecological movements, including 'secular-based' initiatives also provide opportunities for faith members to observe and learn directly or virtually from alternative personal or social practices that help to reduce CO_2 emissions. A breathing space has been created to slow down climate change. However, can this momentum to save planet earth be sustained, as the dominant global economic paradigm continues to shape humanity-planet relationships, and as many peoples, including those in faith communities, remain attached to the consumerist vision of the good life, afflicted by "affluenza" (Hamilton and Denniss 2005) and eschewing the wisdom of "stepping lightly" (Burch 2000), "interbeing" (Thich Nhat Hanh 1998), or "You Are Therefore I am" (Kumar 2002)?

REFERENCES

Alejo, A. E. 2000. *Generating energies in Mount Apo*. Quezon City: Ateneo de Manila University.

Al-Ghazzali Centre. http://alghazzali.org/events/cooks_river (accessed December 12, 2008).

Alliance of Religions and Conservation (ARC). http://www.arcworld.org/about_ARC.htm (accessed January 5, 2009).

Arndt, P. 2005. Catholic Faith and Climate Justice, Chain Reaction # 94, July 2005. http://www.foe.org.au/resources/chain-reaction/editions/94/catholic-faith-and-climate-justice/?searchterm=CatholicFaithandClimateJustice (accessed January 18, 2009).

Artists Project Earth. http://apeuk.org/p083.html (accessed December 22, 2009).

Australian Religious Response to Climate Change (ARRCC). http://www.arrcc.org.au/ (accessed December 29, 2008).

Bagader, A., A. El-Sabbagh, M. Al-Glayand, and M. Samarrai. 2006. Environmental protection in Islam (part 1 of 7): A general introduction. http://www.islamreligion.com/articles/307/ (accessed January 3, 2009).

Barlow, M. 2007. *Blue covenant: The global water crisis and the fight for the right to water*. Toronto: McClelland & Stewart.

Barnes, M. 1994. *An ecology of the spirit: Religious reflection and environmental consciousness.* Lanham, MD: University Press of America.

Barnhill, D. and R. Gottlieb, eds. 2001. *Deep ecology and world religions: New essays on sacred ground.* Albany: State University of New York.

Bernstein, E. 2005. *The splendor of creation: A biblical ecology.* Cleveland, OH: Pilgrim.

Berry, R. J., ed. 2006. *Environmental stewardship: Critical perspectives, past and present.* New York: Continuum.

Berry, T. 1998. *The dream of the earth.* San Francisco: Sierra Books.

Bingham, S. 2009. *Loving God, healing earth: 21 leading religious voices speak out on our sacred responsibility to protect the environment.* Pittsburgh: St. Lynn's Press.

Bullmore, M. A. 1998. The four most important biblical passages for a Christian environmentalism. *Trinity Journal* 19 (141): 139–62.

Burch, M. 2000. *Stepping lightly.* Garbiola Island, BC: New Society.

Burgess, G. 2007. Mosque with a mission. *Greenfutures*, March 9. http://www.forumforthefuture.org/greenfutures/articles/602823 (accessed January 3, 2009).

Catholic Earthcare Australia. http://catholicearthcareoz.net/schools.html (accessed January 22, 2009).

Catholic Schools and Climate Change. http://www.catholicsandclimatechange.org/pdf/K-2lessons.pdf (accessed December 14, 2008).

Cawagas, V. F. 2006. Pedagogical principles in educating for a culture of peace. In *Cultivating wisdom, harvesting peace*, ed. S. H. Toh and V. Cawagas, 299–306. Brisbane: Multi-Faith Centre, Griffith University.

CBCP News. http://www.cbcpnews.com/?q=node/8064 (accessed December 18, 2008).

Chapple, C. K., ed. 2002. *Jainism and ecology: Nonviolence in the web of life.* Cambridge, MA: Harvard University.

Chapple, C. K. and M. E. Tucker, eds. 2000. *Hinduism and ecology. The intersection of earth, sky, and water.* Cambridge, MA: Harvard University.

Chen Xia. 2006. Body in Daoism: An ecological interpretation. Paper presented at the Third International Conference Daoism and Contemporary World: Daoist Cultivation in Theory and Practice, May 25–28, 2006, Abbey Frauenwoerth, Chiemsee, Bavaria, Germany.

China Daoist Ecology Protection Eight Year Plan 2010–2017. http://www.arc-world.org/projects.asp?projectID=382 (accessed January 10, 2009).

Dahl, A. 1995. The Bahá'í approach: Moderation in civilization. Paper read at the Colloquium on Ecology, Ethics, Spiritualities, October 27–29, 1995, Klingenthal, France. http://bahai-library.com/conferences/civilization.html (accessed December 20, 2008).

De Chatel, F. 2003. A pioneer of the environment. *IslamOnline.* http://insideislam.wisc.edu/index.php/radio/eco-islam (accessed January 5, 2009).

Donofrio, J. 2007. A synopsis on greening seminaries. *The Forum on Religion and Ecology Newsletter* 1 (3), November. http://ecologyfellowship.org/2007/11/14/forum-on-religion-ecology-green-seminary-initiative/ (accessed January 3, 2009).

Earth Charter. 2000. Paris: Earth Charter Commission, UNESCO.

Earth Ministry. 2003. The cry of creation: A call for climate justice. http://www.earthministry.org/resources/publications/cry-of-creation (accessed December 13, 2008).

Edwards, D. 2006. *Ecology at the heart of faith.* Maryknoll, NY: Orbis Books.

Ehrenfeld, D. and P. J. Bentley. 1985. Judaism and the practice of stewardship. *Judaism* 34 (3): 301–11.

Faith and the Common Good. http://www.faith-commongood.net/ (accessed December 5, 2008).

Fien, J. 2006. Teaching and learning for a sustainable future. CD ROM. UNESCO. http://www.unesco.org/education/Hsf/ (accessed January 3, 2009).

Fisher-Ogden, D. and S. Saxer. *World religions and clean water laws*. https://www.law.duke.edu/journals/delpf/downloads/delpf17p63.pdf (accessed January 7, 2009).

Gardner, G. T. 2006. *Inspiring progress: Religions' contributions to sustainable development*. New York: Norton.

Garrigues, L. 2008. Kayapo of Brazil protect, preserve land with sustainable agriculture. *Indian Country Today*, September 10.

Gaspar, K. M. 1990. *A people's option: To struggle for creation*. Quezon City: Claretian Publications.

Gelling, P. 2008. Indonesians use Koran to teach environmentalism. *International Herald Tribune*, May 6. http://www.indonesia-ottawa.org/information/details.php?type=news_copy&id=5306 (accessed January 12, 2009).

Girardot, N. J., J. Miller, and L. Xiaogan, eds. 2001. *Daoism and ecology: Ways within a cosmic landscape*. Cambridge, MA: Harvard University Center for the Study of World Religions.

Grim, J., ed. 2001. *Indigenous traditions and ecology: The interbeing of cosmology and community*. Cambridge, MA: Harvard University.

Hamilton, C. and R. Denniss. 2005. *Affluenza: When too much is never enough*. Crows Nest, NSW: Allen & Unwin.

Jackson, L. 2008. Leading sustainable schools: What the research tells us. Goldaming: WWF-UK. http://www.arcworld.org/downloads/14669_lead_sus_school%20(2).pdf (accessed January 22, 2009).

Jones, K. 2003. *The new social face of Buddhism*. Boston: Wisdom Publications.

Kaza, S. and K. Kraft, eds. 2000. *Dharma rain: Sources of Buddhist environmentalism*. Boston: Shambhala.

Khalid, F. M. 2002. Islam and the environment. In *Encyclopedia of global environmental change*. Vol. 5, Social economic dimensions of global environmental change, ed. P. Timmerman, 332–39. Chichester: Wiley.

Knudtson, P. and D. Suzuki. 1992. *Wisdom of the elders*. Toronto: Stoddard.

Kumar, S. 2003. *You are, therefore I am*. Devon, UK: Green Books.

McDonagh, S. 1986. *To care for the earth*. London: Geoffrey Chapman.

Message of the Primates of the Orthodox Churches. October 12, 2008. http://www.ec-patr.org/docdisplay.php?lang=en&id=995&tla=en (accessed December 12, 2008).

Mining Watch. http://www.miningwatch.ca/index.php?/Indigenous_Issues.

Mische, P. 2006. The earth as commodity or community? The wisdom and power of interfaith networks to create a community and culture of life. In *Cultivating wisdom, harvesting peace*, ed. S. H. Toh and V. Cawagas, 155–68. Brisbane: Multi-Faith Centre, Griffith University.

Nelson, L. E., ed. 1998. *Purifying the earthly body of God: Religion and ecology in Hindu India*. New York: State University of New York.

Ozdemir, I. 2003. Toward an understanding of environmental ethics from a Qur'anic perspective. In *Islam and ecology: A bestowed trust*, ed. C. Richard, F. Frederick, M. Denny, and A. Baharuddin, 3–37. Cambridge, MA: Harvard University.

Prime, R. 2006. *Hinduism and ecology: Seeds of truth*. London: Cassell and WWF.

Ruether, R. R. 1992. *Gaia and God*. New York: HarperSanFrancisco.

Sam Aola Ooko. 2008. Green fishing, according to Islam. *EcoWorldly*. http://ecoworldly.com/2008/04/10/green-fishing-according-to-islam/ (accessed January 10, 2009).

Sears, D. *Ecology and spirituality in Jewish Tradition Chabad.Org.* http://www.chabad.org/library/article_cdo/aid/255521/jewish/Ecology-and-Spirituality-in-Jewish-Tradition.htm (accessed January 22, 2009).

Shiva, V. 2002. *Water wars.* Toronto: Between the Lines.

———. 2005. *Earth democracy.* Cambridge, MA: South End Press.

Sikhism and the environment. http://www.sikhroots.net/resources/Sikh_Religion_and_Ecology.pdf (accessed December 9, 2008).

Singh Ji, Bhai Sahib Mohinder, 2008. *Faith and the environment.* http://www.kpsingh.co.uk/blog/2008/06/18/sikh-faith-the-environment/ (accessed December 17, 2008).

Sivaraksa, S. 2005. *Socially engaged Buddhism.* Delhi: B. R. Publishing Corporation.

Sorrell, R. D. 1988. *St. Francis of Assisi and nature: Tradition and innovation in western Christian attitudes toward the environment.* Oxford: Oxford University Press.

Swearer, D. K. 1988. Buddhism and ecology: Challenge and promise. *Earth Ethics* 10 (1): 19–22.

The Shalom Center. http://www.shalomctr.org/node/1363 (accessed December 16, 2009).

Thich Nhat Hanh. 1998. *Interbeing.* Berkeley: Parallax Press.

Tirosh-Samuelson, H., ed. 2003. *Judaism and ecology created world and revealed word.* Cambridge, MA: Harvard University.

Toh Swee-Hin (S. H. Toh) 2002. Citizenship education for a culture of peace: themes, challenges and possibilities. In *Values in education*, ed. S. Pascoe, 38–58. Deakin West, ACT: Australia College of Educators.

Tucker, M. E. and J. Grim. 1994. Overview of world religions and ecology. *Forum on Religion and Ecology.* http://fore.research.yale.edu/religion/ (accessed January 10, 2009).

Tucker, M. E. and D. R. William, eds. 1998. *Buddhism and ecology: The interconnections of dharma and deeds.* Cambridge, MA: Harvard University.

Webber, T. 1989. *Hugging the trees: The story of the Chipko movement.* New York: Penguin.

World Council of Churches. *Justice, peace and creation concerns.* http://www.wcc-coe.org/wcc/what/jpc/ecearth-climatechange.html (accessed January 9, 2009).

World Council of Churches. Climate justice for all. *A statement from the WCC to the High-Level Ministerial Segment of the UN Climate Conference in Nairobi.* http://www.kairoscanada.org/en/ecojustice/climate-change/climate-justice (accessed January 22, 2009).

11 Public Health Threats in a Changing Climate

Meeting the Challenges through Sustainable Health Education

Janet Richardson and Margaret Wade

THE HEALTH EFFECTS OF CLIMATE CHANGE

Climate change presents significant challenges to the ability of populations to maintain health and deliver services to the sick. The impacts are particularly relevant to education for health professionals as the adverse effects are likely to have significant effects on the poor; energy poverty will increase and food security will be compromised, thus increasing inequalities in health (Haines et al. 2006; McMichael et al. 2006). The World Health Organisation (WHO) estimates that the effects of climate change that have occurred since the mid-1970s may have caused over 150,000 deaths in the year 2000 and concludes that these impacts are likely to increase in the future. The impacts of climate change on human health will be unevenly distributed around the world, with developing countries and densely populated coastal areas being particularly vulnerable. However a warming climate may bring some benefits to localized areas; for example a decrease in winter deaths and greater food production at high latitude regions, though this may be compromised by population migration and the associated problems this would bring (see WHO).

The need for countries to undertake national assessments of the impacts of climate change on human health was noted as early as 1999 (Kovats et al. 1999). McMichael et al. (2006) review the published estimates of future health effects of climate change and present a schematic summary of the main pathways by which climate change affects health. Significantly the authors differentiate between climate–health relationships that are easy to define (physical hazards due to heat waves, flooding, storms, fires, and infectious disease), and those that are less easy to quantify but may have detrimental effects including risks of malnutrition and mental health problems (disruption to regional food supplies and fisheries, population displacement, loss of livelihoods). Earlier assessments of the potential effects of climate change show similar pathways and emphasize the need for mitigation to reduce CO_2 emissions and energy use, and adaptation measures to address the consequences (Haines and Patz 2004; Haines et al. 2006).

Clean water supplies, food security, shelter, and secure communities play a significant role in health maintenance and disease control. Specific effects, challenges, and possible solutions can be illustrated by using the United Kingdom (UK) as an example. A report (Department of Health and Health Protection Agency 2008) sets out the potential health impacts of climate change: in the UK an increase in mean annual temperature of 2.5– 3°C by the end of this century is likely to have a number of consequences. Increases in flooding will result in loss of life, injuries, water contamination and restrictions, loss of homes resulting in temporary accommodation, and potential problems with businesses and employment. Rising water temperature is likely to lead to increases in algal blooms in reservoirs with consequent decreased efficiency in chemical coagulation increasing the potential for polluted drinking water. Excessive rainfall will lead to an increased number of pathogenic bacteria in surface water due to run-off from agricultural land. Increased heat will result in drought and disruption of water supplies, and frequent and long heat waves resulting in dehydration and deaths. Higher levels of air pollution will lead to greater respiratory problems, increases in skin cancers, and an increase in the incidence of food-borne diseases. Though malaria is unlikely to become a problem in the UK, vector-borne diseases such as Lyme disease[1] may become more prevalent.

The consequences will include higher demands on emergency and health services (for example those required to cope with flooding and infectious disease outbreaks), and rising mental health problems (due to displacement, homelessness, and possible loss of livelihoods). Haines et al. (2006) emphasize the need for public health strategies such as improved surveillance in order to adapt to the impact of climate change, as well as advocating action to mitigate climate change by reducing the use of fossil fuels.

HEALTHY PUBLIC POLICY AND HEALTH EDUCATION

Public Health Policy

What can health care providers and educators do about this and how can public health and health promotion support individuals and communities through mitigation and adaptation measures? Future public health planning will need to take account of predicted possible scenarios and will inevitably involve multi-agency working and detailed assessment of local resources.

One initiative aimed at enabling people and organizations in the UK National Health Service (NHS) and wider community to engage in activities that support sustainability is 'The Convergence of Health and Sustainable Development' network. This is a network of health and other professionals committed to promoting sustainable development in the NHS and wider community and strengthening the position of sustainable development within the NHS workforce.

The Climate and Health Council has been established as a not-for-profit international organization aimed at mobilizing health professionals across the world to take action to limit climate change and its effects on human health (see http://www.climateandhealth.org/). A recently published document outlines the action that can be taken at organizational and individual levels (Faculty of Public Health 2008). This was closely followed by the publication of the *The Health Impact of Climate Change: Promoting Sustainable Communities—Guidance Document* (Department of Health 2008). While the potential effects of climate change and energy efficiency are being considered by a number of public health agencies in the UK, limited information is available about current practical efforts to implement mitigation and adaptation measures (Richardson et al. 2008). The *NHS Good Corporate Citizenship Guidance* on the Sustainable Development Commission Web site[2] describes how NHS organizations can embrace sustainable development and tackle health inequalities through their day-to-day activities. This self-assessment model helps organizations to identify and assess their contribution to good corporate citizenship and suggests ideas for future action, providing guidance on transport, procurement, facilities, management, and new buildings. Examples of action include sourcing healthy and locally produced food, and the development of Park and Ride schemes (facilities for people to drive their cars to large parking areas on the edge of towns and then take the bus into the town center, thus reducing the number of cars in town centers). More recently the UK NHS has launched a *Carbon Reduction Strategy* that sets out a framework for NHS organizations in England to reduce their carbon emissions (NHS Sustainable Development Unit 2009).

Health Education

It has been suggested that the responsibility of healthcare practitioners to protect and promote the health of the public should be extended to working to prevent climate change (Gill et al. 2007). Traditionally, health professionals have largely ignored environmental, social, economic, and political structures and focused almost exclusively on individual risk factors related to diseases. Yet the five environmental characteristics, those of pure air, pure water, efficient drainage, cleanliness, and light, held by Florence Nightingale to be essential for promoting health, are as important to nursing today in the twenty-first century as they were in the 1800s (Leddy 2006). Nurses need to be encouraged to adopt a broader agenda than a narrow biomedical framework, acknowledge the sociopolitical determinants of health, and through a healthy public policy framework contribute to the creation of supportive environments (Wills 2007). Contemporary public health needs to recognize the interrelatedness of health and social inequalities and health-sustaining environments in order to protect the most vulnerable from global environmental changes such as heat-related illnesses (McMichael and Beaglehole 2000).

Health education is an essential building block of healthy public policy (Tones and Green 2004). The purpose of healthy public policy and health education should be to empower individuals and communities and to remove or reduce the barriers to the achievement of health for everyone (Tones 2001). By creating knowledge, health education can lead to critical consciousness-raising and empower individuals and communities to influence healthier policies through public pressure, networking, and lobbying (Douglas and Jones 2007).

Health education can take place at the three different levels of primary, secondary, and tertiary prevention. Primary prevention is concerned with seeking to avoid the onset of ill health through the provision of advice and counseling, secondary prevention through providing education to shorten the progression of ill health, and tertiary prevention seeking to limit disability or complications of irreversible conditions (Naidoo and Wills 2005). Mitigating global climate change not only involves government action but also requires voluntary cooperation from the public through raising awareness and supporting behavioral change related to climate change (Semenza et al. 2008). Primary preventive education of individuals and communities on the benefits of active transport, including alternatives such as walking and cycling, may contribute to a reduction in air pollution. Education on healthy eating, increasing the consumption of fruit and vegetables with less reliance on red meat, will reduce the level of carbon dioxide through livestock production (Frumkin and McMichael 2008). Therefore, encouraging healthy lifestyles, such as healthy eating and promoting walking and cycling will not only benefit the individual's health, for example reducing the risks of coronary heart disease, type 2 diabetes, and colorectal cancer, but also address wider environmental sustainability issues. Health practitioners need to promote the co-benefits of leading healthy lifestyles and the positive impact this may have on the environment.

THE ROLE OF HEALTH PRACTITIONERS AND EDUCATORS

According to Frumkin and McMichael (2008), the timescale for anticipating future health needs is unprecedented, moving far beyond the traditional boundaries of health planning. Thus the health sector has a responsibility for developing public health and health-promotion strategies to address the serious impact on health from climate change for decades to come. Primary Care Trusts (PCTs)[3] are charged with improving the health of local populations and leading the development of health-improvement plans to maximize public health activity at a local and community level (Peckham and Taylor 2003). Now and in the future, health professionals will be required to make best use of their knowledge and skills to develop primary prevention measures to mitigate, as well as secondary prevention measures to adapt to, climate change, in order to reduce climate-related disease, injury, and death (WHO 2008).

Health professionals need to ensure that they are competent in primary and secondary prevention strategies in order to provide clear advice and information to the public on the effects of climate change and the risks posed to health. As well as health practitioners being equipped with evidence-based information on the threats posed to health from climate change, it is also essential that they understand the principles of effective health education, supporting behavior change and the impact of economic, social, and environmental factors on health and wellbeing.

Contemporary Public Health

The World Health Organisation through the *Declaration of Alma Ata* (WHO 1978) was instrumental in broadening the concept of health and disease by adopting a holistic view of health stating that "health, which is a state of complete physical, mental and social wellbeing, and not merely the absence of disease or infirmity, is a fundamental human right" (WHO 1978, 1). In recognizing the wider determinants of health and disease, the *Declaration* emphasized a move away from the narrow focus of individual to community responsibility.

In the 1980s there was a major shift in public health policy with an acknowledgment of a wider ecological perspective on health and the emergence of 'New Public Health' (Ashton and Seymour 1988), providing an opportunity to promote health in the context of the challenges that climate change will bring. The World Health Organisation has been fundamental in emphasizing the importance of healthy public policy and adopting the settings approach (health is partly a consequence of the environments in which people live). This has been demonstrated in, for example, the Healthy Cities project, through the *Ottawa Charter for Health Promotion, Health for All* (WHO 1986) and the *Health 21* movement (WHO 1998). The aim of *Health for All*, later updated to *Health 21*, is to promote and protect people's health and to achieve the full potential of all citizens based on the belief that health is a fundamental right (Earle 2007). The *Charter for Public Health* recognizes that all people, in a world of limited resources, have a right to certain basic needs (Public Health Alliance 1987). This right to basic needs will be compromised by climate change rendering many areas of the globe uninhabitable, and will thus produce pressure on resources elsewhere that will adversely affect health.

Public health goals, such as access to a clean and healthy environment are, in the UK and elsewhere, a citizen's statutory right and need to be prioritized through sustainable development to achieve environmental and human health (Secrett and Bullock 2002). Healthy public policy is about creating a healthy society that addresses not only the physical but also social, cultural, and economic aspects of health and wellbeing (Douglas and Jones

2007). For example one of the major challenges facing the Healthy Cities project is the mitigation and adaptation to climate change along with the growing number of people living beyond eighty years (Lawrence and Fudge 2007). Populations most at risk from heat-related illnesses are those living in urban areas without access to adequate air conditioning, with the very old and very young being most at risk. Other risk factors include living in poverty and social isolation (McGeehin and Mirabelli 2001).

Health Education and Empowerment

Individuals, families, and communities require evidence-based information on the effects of climate change in order to make informed decisions and to facilitate an attitude of positive engagement (Frumkin and McMichael 2008). Traditionally, public health evidence has been informed by epidemiology and quantitative research methods. However there is now an acceptance that a qualitative approach is also necessary to explore social determinants that affect people's behavior. An evidence-base of lay knowledge is needed to ascertain clients' preferences, views, and beliefs (Reading 2008), and to engage different age groups in making the links between living sustainably and staying healthy. For example involving children in school garden projects provides an opportunity to combine education about food (fresh and healthy fruit and vegetables, food miles, food security, and production), wildlife, and exercise, while encouraging their appreciation about their own location within and responsibilities for the natural environment.

According to Tones and Green (2004) it is through the process of communicating, informing, and educating individuals and communities that people gain control over their own health. Health education is providing people with information in order to develop knowledge and skills so that they are enabled to make healthy choices. In an unprecedented state of transition brought about by climate change more imaginative approaches will be required, and may even appear in a form of self-organization. Examples can be found following disasters such as hurricanes, where local people rebuild communities, acquiring skills and knowledge and developing new methods of food growing, and leading to greater resilience within the community (Tidball and Krasny 2007).

The fundamental principle of health education is empowerment, where individuals and communities are enabled to make healthy choices for themselves as opposed to expert-led persuasion. For people to be empowered they not only need to be motivated to change their situation but also need the information, support, and skills to do so (Naidoo and Wills 2005). Health education that focuses on empowerment differs dramatically from the traditional model in that instead of adopting a victim-blaming approach based on persuading people to adopt preventative behaviors and to comply with the professional's perspective, the aim of an empowerment model is to strengthen the individuals' capabilities and choice of action (Tones 2001).

The problems of climate change may appear so overwhelming that individuals may feel that they are powerless, and that any action they take may be of no consequence. The challenge for educators is to combine actions towards mitigating and adapting to climate change with the potential (and more immediate) health benefits for individuals. One obvious action is the reduction of car use and increase in physical activity. Combining this with walks in the natural environment encourages a greater connection with the natural world and can promote wellbeing (Milligan et al. 2004).

Facilitating Behavioral Change

Engaging and supporting individuals to adopt healthy lifestyles is complex and requires an awareness of behavioral change processes as well as an understanding of social, psychological, and environmental issues involved in translating behavioral intention into action (Tones and Green 2004). Although individuals may possess knowledge about the damaging effects of climate change and the importance of low-carbon sustainable healthy lifestyles, this does not necessarily equate to a change in behavior. Prochaska and DiClemente's (1984) transtheoretical model describes behavioral change as a process and identifies five sequential stages that individuals may go through in order to achieve sustained behavior change:

- precontemplation (not interested in changing behavior)

- contemplation (understands the reasons to change but no action considered)

- preparation (ready to make a change but not acted upon it)

- action (makes an attempt to change but requires support)

- maintevnance (maintains action)

The individual then either continues to maintain the behavior change or relapses and may go through the cycle again. However, relapse or setback should not be viewed as failure. Although individuals may go backwards and forwards through the cycle, each small change brings benefits in developing a sense of achievement (Naidoo and Wills 2000).

Supporting behavioral change involves encouraging self-determination and promoting self-esteem (sense of personal worth) where individuals recognize for themselves a need to change and become self-motivated. The traditional approach of giving advice with the expectation of initiating change has been criticized as being largely ineffective and of undervaluing the personal costs and situation of the individual (Britt et al. 2004). Furthermore coercion may lead to change but is far more likely to lead to alienation. A more effective approach is to use the person's own values and goals to achieve behavioral change. Adopting a client-centered approach of being supportive, affirmative, and nonjudgmental is an effective way of helping people to consider and plan for change (Tomkins and Collins 2005). Nurses are now encouraged to adopt a client-centered approach and involve clients in their own care. Developing clients' personal skills and supporting confidence-building will enable clients to access and make best use of the health care system, assess their own health risks and decision-making about their lifestyles, as well as enabling them to understand the wider determinants of economic, social, and environmental factors affecting their health (Wills 2007).

Motivational interviewing, developed by Rollnick and Millner Miller, is a client-centered approach that is now being adopted in health care to support health professionals as a way of working in partnership with clients (Britt et al. 2004). Using counseling skills, the practitioner facilitates the individual to explore and resolve ambivalence about behavior change. The key principles of motivational interviewing are expressing empathy, developing discrepancy, avoiding argument, rolling with resistance, and supporting self-efficacy (Rollnick and Millner Miller 1995). In motivational interviewing, resistance to change is acknowledged, but rather than using direct persuasion, the practitioner adopts a partnership approach to explore discrepancies between the client's present behavior and their personal goals. By increasing the belief in their ability to change (self-efficacy), clients are enabled to find solutions for themselves (Britt et al. 2004). This approach can be used to challenge behaviors that are not adaptive to climate change and have direct health consequences.

Translating behavioral intention into action is dependent upon a range of factors. The health action model developed by Tones (Tones and Green 2004) provides a comprehensive framework for health professionals to examine the psychological as well as the social and environmental influences on adopting and maintaining a healthy lifestyle. The model explains that intention to adopt particular health-related behavior is likely to be determined by individual beliefs, motivation, normative influences such as social norms and cultural values, and the self (Tones and Green 2004). High self-esteem, an internal locus of control (a belief that results are primarily from one's own behavior and actions as opposed to external influences), and a positive self-concept are more likely to lead to positive health-related behaviors (Goodman-Brown and Gottwald 2008). The model also recognizes that translating intention into action requires facilitation of knowledge and skills as well as the availability of a supportive environment (Tones and Green 2004).

Voluntary mitigation to adopt low-carbon health behaviors can be maintained if healthy choices are made easier. Therefore, for mitigation of climate change to be effective, as well as behavioral change of individuals, structural changes are also required in order to aid the adoption of healthy behaviors. For example legislative and regulative measures to enhance cycle paths, and school fruit and vegetable schemes (Semenza et al. 2008).

CHANGING PRACTICE FOR A HEALTHY FUTURE

The development of strategy and policy requires governments to take a leading role and to be committed to their implementation. Effective action, however, presents a greater challenge, as this requires behavioral and cultural change for both organizations and individuals. Action needs to be embedded within the culture if it is to be sustainable, so that it becomes second nature. One of the difficulties, which is also inherent in health education/promotion, is that the health outcomes of positive behavioral change are not immediate and, in the case of climate change, may be perceived to be of greater benefit to future generations rather than to the individual who is expected to change their behavior. For example a reduction in car use may reduce CO_2 emissions and have long-term consequences for the environment. However it will also have more immediate effects by reducing airborne toxins (that can cause respiratory disease), reducing road traffic accidents, and increasing physical activity.

The traditional approach of the environmental movement that uses a combination of fear and information fails to engage the public (Hounsham 2006). This approach to health education also has its limitations (Frumkin and McMichael 2008). Rather, an approach that connects with peoples' values, desires, emotions, and presents positive visions within their sphere of influence through their ability to engage, is likely to have more success. Learning is central to this. In his exploration of an ecological worldview Sterling (2007) emphasizes a "relational" or "participative" view that requires at least second-order social learning[4]: "fundamental questioning and re-ordering of assumptions." Sterling suggests that social learning agents need to "balance security and challenge in learning situations to facilitate deep questioning, creativity and innovation" (74), and offers several ways in which this can be achieved. In the next section examples are provided of grassroots actions that support and promote local engagement, and policy directives that foster health and sustainability. We will use these examples to illustrate Sterling's approach, in order to emphasize how such actions could be extended to reflect second-order (meta-cognitive) social learning.

Richardson et al. (unpublished) sought examples of good practice of the systematic effects and consequences of proposals, policies, and actions (Sterling 2007) in relation to climate change policies by assessing publically available data on the Web sites of NHS Primary Care Trusts in England. Examples of action fell into the following categories: strategic initiatives, joint working with other agencies, promoting sustainable communities, and targeted actions. These examples were clearly linked to Government policy and thus were grounded in the currently accepted framework.

In this UK study these examples demonstrate an organizational commitment towards action to mitigate and adapt to climate change and energy vulnerability. However the specific actions taken and the consequences of policy and strategy documents were not apparent from the details available via the Web sites. This policy–practice gap is likely to be in evidence in other countries and worthy of further investigation (Richardson et al. 2008). Furthermore, there was no evidence that the policies brought about fundamental changes to the system, or had changed the worldviews of those within the system.

The interdependence of health and food security is important for building sustainable and resilient communities and provides a social learning opportunity for future coping. Following the decline in homegrown food, there is now a need for sustainable food systems such as gardening and urban agriculture (Lang 2008). *Community-based activities that involve participative action on community-led initiatives* (Sterling 2007) include promoting food growing and access to green spaces. This can help communities build resilience and social capital, thus enabling them to better deal with disasters such as those brought about by climate change (Tidball and Krasny 2007). The concept of resilience and psychological hardiness is associated with better physical health and refers to an individual's capacity to withstand stressors (Lambert and Lambert 1993). Resilience can be enhanced through outdoor education, leading to increased hardiness, self-confidence, and self-efficacy (Hattie et al. 1997). The physical and psychological benefits of gardening and horticulture are well documented, and projects supporting gardening as a therapeutic activity are on the increase. Horticulture therapy has been shown to increase interpersonal and motivational skills, punctuality, health and fitness, and enhance psychological wellbeing, promote social inclusion and a sense of place and belonging, thus reducing social isolation; such initiatives provide a structure and daily routine for people who face health and wellbeing challenges, and can equip them with new skills and provide opportunities for social interaction (Midden and Barnicle, unpublished; Sempick and Aldridge, unpublished; Milligan et al. 2004).

Gardening projects enable people to engage in an activity which may be otherwise beyond their means or circumstances. There are obvious co-benefits of linking outdoor physical activity through gardening with information about food, for example, increased knowledge of healthy diet, enhanced fitness, and community social cohesion. Such initiatives may be based within the currently accepted framework, but have the capacity to fundamentally influence the worldview of those who take part, and provide an important opportunity to engage individuals in building skills that support resilience at times of turmoil and transition.

One new social movement is taking place that attempts to develop community-led action to limit the effects of peak-oil and climate change, and to enable communities to build resilience in order that communities can respond effectively to the challenges. This *visioning of alternative futures*

(Sterling 2007) is now a worldwide phenomenon with more than 100 towns or cities developing transition initiatives (Transition Town Network). The movement started with Transition Town Totnes[5] (TTT). A key concept is resilience, both communal and personal, in response to change. Transition Initiatives are based on four key assumptions (Hopkins 2008): that life with dramatically lower energy consumption is inevitable, and needs to be planned for; that communities presently lack the resilience, and often necessary skills, to respond appropriately; that we have to act collectively and now; that we need to unleash our creativity so that we can develop more connected and enriching ways of living, in harmony with human development and the earth. This approach provides an opportunity for local people to connect different aspects of health and wellbeing together through integrated projects (Transition Town Health and Wellbeing Group); it challenges the consensually accepted order and is driven by groups and individuals with radical worldviews based on relational and ecological perspectives.

There are a number of ways to *use learning opportunities* to embed sustainability into healthcare curriculum. For example nurses will need to acquire knowledge and skills to address public health needs to improve health and reduce inequalities of health. Therefore it is imperative that sustainability, as a major public health issue, becomes embedded into the curriculum. Emphasizing the more general co-benefits of sustainable development and health is crucial to this process, rather than simply focusing on those areas (such as public health) that have a more obvious relationship with sustainability issues. For example in addressing areas such as 'the child in the community' and self-care, emphasis can be placed on the co-benefits of climate change mitigation activity (such as decreased car use/increased walking) and benefits to health (through physical activity). Education for health and climate change can be embedded throughout the school curriculum in formal teaching settings. For example by taking climate change as the underpinning topic, disciplines such as environmental science, biology, geography can explore the consequences of changes on communities, their livelihoods and health, bringing in issues regarding mass migration, poverty, food security and farming, infectious disease, and so on. Informal education linking health with the environment can take place through activities such as Forest School[6] and the greening/landscaping of school play areas.

Innovative approaches to education provide opportunities to mitigate climate change. Live interactive Webcasting can reduce unnecessary travel by students and staff while offering more interactivity for students and the possibility of internationalizing their experience. For example health and social work students at the University of Plymouth, UK, in particular are widely distributed throughout the South West Peninsula of England requiring long-distance travel to lectures. The delivery of, for example, just one lecture for thirty-four midwifery students replaced by a Webcast saved 3672 miles of (mostly car) travel.[7] This method of delivering education could be particularly relevant to rural communities throughout the world.

CONCLUSIONS: EMBEDDING SUSTAINABILITY INTO THE HEALTH EDUCATION AGENDA

Formal and informal settings can provide opportunities for learning about health and climate change. In particular, opportunities that promote the positive health benefits of living sustainably are likely to have greater impact and bring about behavior change. In schools, outdoor learning can encourage children to have a positive and respectful approach to the natural environment, and to investigate the health promoting properties of plants and the natural world. This can be combined with education about food and exercise.

Learning can take place through focused action to develop resilience and empower and enable vulnerable people to adapt to climate change through investments in social protection, health, education, and other measures. Initiatives such as the Transition Town movement that aim to build resilience in order to adapt to a low-carbon economy and reduced availability of oil can also promote adaptation to climate change (see Transition Town Network Web site). At times of unprecedented transition, social learning situated within the context of building resilience has the potential to stimulate the personal and social adaptation required to meet new challenges.

Education of health care agencies is required for them to implement mitigation measures to reduce CO_2 emissions and energy use (Richardson et al. 2008). For example, procurement policies for energy-efficient buildings, reducing travel by promoting tele/Web- and videoconferencing, implementing green travel plans, and recycling. Scenarios can be used to develop joint working and multiagency strategies to proactively promote sustainability and meet the needs of emergencies arising from the impact of climate change and energy vulnerability.

Emphasizing and making explicit the co-benefits of sustainability provides a learning opportunity for individuals and organizations to see the potential immediate benefits of what might usually be viewed as an initiative with only long-term outcomes. For example the co-benefits to organizations of carbon-reduction strategies can lead to early savings on fuel costs and local control over supply; the introduction of green travel plans can lead to a decrease in traffic-related deaths (Roberts and Hillman 2008). The co-benefits to individuals of reduced car use include greater physical activity and the associated health benefits this brings, involvement in local food growing initiatives leading to greater social cohesion, increased mental health and wellbeing, and physical benefits.

The key learning points for the health professions are based on public health as a wider concept than simply health and disease. Public health is as much about the built and natural environment as it is about health education. This requires a visionary and proactive approach that engages communities and incorporates multidisciplinary and multiagency partnership working in order to construct and plan for scenarios involving climate change. Education and learning needs to be embedded in the process of training health practitioners. Ultimately, health care educators will need the skills to change their own behavior as well as support the necessary changes in others. Thus 'walking the talk' and modeling the very behavior they are attempting to promote!

SCENARIOS

Lynas (2007) presents a sobering picture of the consequences of an average global temperature increase of up to six degrees; a rise of five degrees will result in "zones of uninhabitability" where large-scale developed society is no longer sustainable (224).

It is impossible to predict precisely what the effects of global warming will be on specific regions. In the UK, for example, climate change scenarios have been constructed in order to assess the potential health impacts of different rises in temperature (Department of Health and Health Protection Agency 2008). Two scenarios and snapshots follow based on small and larger temperature increases and different mitigation and adaptation approaches. Both report the action of public health agencies based on prior learning and motivation in response to climate change. The focus is on public health agencies rather than individuals as these agencies will be instrumental in planning for and managing health and emergency responses in times of turmoil and disaster.

Scenario 1: Setting—England 2020

Average winter temperatures have increased by 1.5°C and summer temperatures by 2.5°C. Average winter rainfall has increased by 15 percent and summer rainfall reduced by 20 percent. Storms, rising sea levels, coastal flooding with storm surges are more common. A combination of increased rainfall and high tides has caused prolonged flooding in a town in South West England. Emergency services are still working hard seven days into continuous rain, and the water levels are still rising. As the town is surrounded by agricultural land there is a risk of contamination of water by chemicals and bacteria and water supplies are under threat. This risk is increased by the excessive demand on sewerage and drainage facilities which are unable to cope, thus there is a danger of increased risk of infectious diseases. Transport is hampered and food supplies are compromised.

Snapshot

As a consequence of *investment in educational infrastructure* promoting and supporting adaptive responses to climate change, central and local government policy has been quick to respond to new evidence of more rapid increases in global warming than was previously predicted. A Web-based Climate Change Response initiative links intelligence from the World Health Organisation and the new International Meterological Office (providing short- and long-term weather predictions) to a distributed network of emergency services and support organizations worldwide. Mobile phone 'text messages' are relayed to individuals in areas under threat. These systems are embedded in education throughout the school and university curricula, providing students with opportunities to study mitigation and adaption effects in real-time, and to develop enhanced solutions (thus supporting creativity as a response to challenges and threats). An enhanced Health Protection Agency, working closely with the Environment Agency activated emergency planning rapidly in response to predictions of adverse weather. Clean water supplies were secured prior to the flooding and emergency food supplies delivered to areas likely to be affected. The food-rationing policies, significant local food growing (with more people involved in growing and harvesting), and local distribution controls has resulted in a generally more healthy and physically active population. As a consequence of public health involvement in supporting and building 'sustainable communities' through local public engagement and capacity building for individual learning (Glasser 2007), 'Education for Resilience' is a core-curriculum subject and resilient community networks are in place that ensure those at risk are supported by those who are more able. Compulsory 'Gardening Leave' within all private and public organizations has provided additional skills-building for food growing and emergency support. These networks respond to this incident by working closely with the emergency services to provide temporary accommodation for those who need it.

SCENARIO 2: SETTING—ENGLAND 2080

Average summer temperatures have increased by 4.5°C and rainfall reduced by 40 percent. Zones of uninhabitability have resulted in global mass migration, malnutrition, starvation, and conflict. The coastal areas have already seen villages and coastal towns disappear under water, and those still standing are constantly under siege from refugees seeking habitable environments. The motivation to address climate change was compromised by successive governments that failed to learn from significant hurricane events and were predominantly concerned with the possibility that refocusing lifestyle away from consumerism would result in loss of votes. Therefore mitigation and adaptation measures were implemented too late to ensure a smooth transition in the face of the changing climate, thus requiring the rapid implementation of a number of radical policies on population growth and immigration between 2030 and 2040.

Snapshot

It is 2080 and the public health services are being stretched as they attempt to cope with the ongoing health problems brought about by changes in the world's climate. The increase in water- and vector-borne diseases has resulted in an expansion of the Health Protection Agency. There is now significant health surveillance including reporting of disease outbreaks and increased water testing, but even this fails to deal effectively with the spread of infection, resulting in a rapid rise in the number of deaths from infectious diseases and food poisoning. The consequences of the controversial population control measures implemented in early 2040 are now evident. The 'One Child Policy' has resulted in a reduction in the UK birth rate. Although this policy remains contentious, the benefits of population control within diminishing resources and unsustainable life-styles has now become all too obvious. The more radical implementation of the 'Health Home Guard' and border control policies that would once have seemed unimaginable are now an acceptable form of sustainability practice. Only a limited number of refugees from uninhabitable lands can be accommodated, those who cannot be accommodated are turned away at the borders and have nowhere else to go. Many die of starvation on rafts in the open waters surrounding the UK, or are shot trying to gain entry illegally. This requires the Health Home Guard to be fully engaged with the Border Control Authority in order to keep the waters free from decaying bodies.

NOTES

1. Lyme disease is a vector-borne disease carried by ticks; symptoms vary from flu-like symptoms to joint pain (http://www.lymediseaseassociation. org/).
2. http://www.sd-commission.org.uk/presslist.php/1/helping-the-nhs-to-be-a-good-corporate-citizen.
3. PCTs are responsible for commissioning a comprehensive and equitable range of services, within allocated resources, across all health service sectors. They have significant resources and the potential to influence tendering/contracting processes and procurement activities.
4. First order social learning—basic learning within a consensually accepted framework. Third order social learning—leads to a complete change of worldview (Sterling 2007).
5. TTT acts as a catalyst for Totnes (Devon, UK) and district to explore how to prepare for and adapt to life beyond cheap oil and gas, and respond proactively to climate change.
6. Forest Schools are an innovative approach to outdoor education where participants learn an appreciation of the natural environment: http://www.forestschools.com/.
7. Professor R. Jones, University of Plymouth, personal communication. http://www.plymouth.ac.uk/health/webcasts

REFERENCES

Ashton, J. and H. Seymour. 1988. *The new public health.* Buckingham, UK: Open University Press.

Britt, E., S. M. Hudson, and N. M. Blampied. 2004 Motivational interviewing in healthy settings: A review. *Patient Education and Counseling* 53: 147–55.

Climate and Health Council. http://www.climateandhealth.org/ (accessed June 26, 2009).

Department of Health. 2008. *The health impact of climate change: Promoting sustainable communities.* http://www.dh.gov.uk/en/Publicationsandstatistics/Publications/PublicationsPolicyAndGuidance/DH_082690. (accessed June 26, 2009).

Department of Health and Health Protection Agency. 2008. *Health effects of climate change in the UK: An update of the Department of Health Report 2001/2002.* http://www.dh.gov.uk/en/Publicationsandstatistics/Publications/PublicationsPolicyAndGuidance/DH_080702 (accessed June 26, 2009).

Douglas, J. and L. Jones. 2007. The development of healthy public policy. In *Policy and practice in promoting public health,* ed. C. Lloyd, S. Handsley, J. Douglas, S. Earle, and S. Spurr, 33–64 London: Sage Publications.

Earle, S. 2007. Promoting public health in a global context. In *Policy and practice in promoting public health,* ed. C. Lloyd, S. Handsley, J. Douglas, S. Earle and S. Spurr, 1–32. London: Sage Publications.

Faculty of Public Health. 2008. *Sustaining a healthy future: Taking action on climate change.* http://www.fphm.org.uk/resources/AtoZ/r_suataining_a_healthy_future.pdf

Frumkin, H. and A. J. McMichael. 2008. Climate change and public health: Thinking, communicating and acting. *American Journal of Preventive Medicine* 35 (5): 403–10.

Gill, M., F. Godlee, R. Horton, and R. Stott. 2007. Doctors and climate change. *British Medical Journal* 1 (335): 1104–05

Goodman-Brown, J. and M. Gottwald, eds. 2008. *Public health interventions.* In *Public health approaches to practice,* ed. J. Mitcheson, 94–115. Cheltenham: Nelson Thornes.

Glasser, H. 2007. Minding the gap: The role of social learning in linking our stated desire for a more sustainable world to our everyday actions. In *Social Learning: towards a sustainable world,* ed. A. Wals, 35–61. Wageningen, The Netherlands: Wageningen Academic.

Haines, A., R. S. Kovats, D. Campbell-Lendrum, and C. Corvalan. 2006. Climate change and human health: Impacts, vulnerability and public health. *Journal of the Royal Institute of Public Health* 120: 58596.

Haines, A. and J. Patz. 2004. Health effects of climate change. *Journal of the American Medical Association* 291: 99–103.

Hattie, J., H. W. Marsh, J. T. Neill, and G. E. Richards. 1997. Adventure education and outward bound: Out-of-class experiences that make a lasting difference. *Review of Educational Research* 67: 43–87.

Hopkins, R. 2008. *Transition handbook.* Dartington, UK: Green Books.

Hounsham, S. 2006. *Painting the town green.* London: Green Engage.

Kovats R. S., H. Haines, R. Stanwell-Smith, P. Martens, B. Menne, and R. Bertollini 1999. Education and debate: Climate change and human health in Europe *British Medical Journal* 318: 1682–85

Lambert, C. and V. A. Lambert. 1993. Relationships among faculty practice involvement, perception of role stress, and psychological hardiness of nurse educators. *Journal of Nurse Education* 32 (4):171–79.

Lang, T. 2008. A food crisis is heading our way. http://www.guardian.co.uk/commentisfree/2008/oct/16/food-agriculture (accessed July 7, 2009).

Lawrence, R. and C. Fudge. 2007. Healthy cities: Key principles for professional practice. In *Public health: Social context and action*, ed. A. Scriven and S. Garman, 180–92. Maidenhead, UK: Open University Press.

Leddy, S. K. 2006. *Integrative health promotion: Conceptual basis for nursing practice*. 2nd ed. London: Jones and Bartlett.

Lynas, M. 2007 *Six Degrees: Our future on a hotter planet*. London: HarperCollins.

McGeehin, M. A. and M. Mirabelli. 2001. The potential impacts of climate variability and change on temperature related morbidity and mortality in the United States. *Environmental Health Perspectives* 109 (2).

McMichael, A. J. and R. Beaglehole. 2000. The changing global context of public health. *The Lancet* 356: 495–99.

McMichael, A. J., R. E. Woodruff, and S. Hales. 2006. Climate change and human health: Present and future risks. *The Lancet* 367: 859–69.

Midden, K. S. and T. Barnicle. *Evaluating the effects of a horticulture program on the psychological well-being of older persons in long-term care*. http://www.actahort.org (accessed June 26, 2009).

Milligan, C., A. Gatrell, and A. Bingley. 2004. 'Cultivating health': Therapeutic landscapes and older people in northern England. *Social Science and Medicine* 58:1781–93.

Naidoo, J. and J. Wills. 2005. *Public health and health promotion: Developing practice*. Edinburgh: Ballière Tindall.

Naidoo, J. and J. Wills. 2000 *Health promotion: Foundations for practice*. 2nd ed. London, Ballière Tindall.

NHS Sustainable Development Unit. *Saving carbon, improving health: NHS carbon reduction strategy* 2009. http://www.sdu.nhs.uk/page.php?area_id=2 (accessed June 26, 2009).

Peckham, S. and P. Taylor. 2003. Public health and primary care. In *Public health for the 21st century*, ed. J. Orme. J. Powell, P. Taylor, T. Harrison, and M. Grey, 93–106. Maidenhead, UK: Open University Press.

Prochaska, J. O. and C. C. DiClemente. 1984. *The transtheoretical approach: crossing traditional boundaries of therapy*. Homewood, IL: Dow Jones Irwin.

Public Health Alliance. 1987. *The essentials for every citizen's right to good health*. http:// www.wherts-pct.nhs.uk.

Reading, S. 2008. Research and development: Analysis and interpretation of evidence. In *Public health skills*, ed. L. Coles and E. Porter, 170–84. Oxford: Blackwell.

Richardson, J., F. Kagawa, and A. Nichols. Unpublished. Health, energy vulnerability and climate change: A retrospective thematic analysis of Primary Care Trust policies and practices.

Richardson, J., F. Kagawa, and A. Nichols. 2008. Health, climate change and energy vulnerability: A retrospective assessment of Strategic Health Authority policy and practice in England. *Environmental Health Insights* 2: 97–103.

Roberts, I. and M. Hillman. 2008. *Climate change: The implications for health policy on injury control and health promotion*. http://www.injuryprevention.bmj.com (accessed November 10, 2008).

Rollnick, S. R. and W. R. Millner Miller. 1995. What is motivational interviewing? *Behavioural Cognitive Psychology* 23: 325–34.

Secrett, C. and S. Bullock. 2002. Sustainable development and health. In *Promoting health*, ed. L. Adams, M. Amos, and J. Mauno, 34–46. London: Sage Publications.

Semenza, J. C., D. E. Hall, D. J. Wilson, B. D. Bontempo, D. J. Sailor, and L. A. George. 2008. Public perception of climate change: Voluntary mitigation and barriers to behaviour change. *American Journal of Preventive Medicine* 35 (5): 479–87.

Sempick, J. and J. Aldridge. Unpublished. *Social and therapeutic horticulture in the UK: The growing together study.* Leicestershire, UK: Loughborough University.

Sterling, S. 2007. Riding the storm: Towards a connective cultural consciousness. In *Social Learning: towards a sustainable world,* ed. A. Wals, 63–82. Wageningen, The Netherlands: Wageningen Academic.

Tidball, K. G. and M. Krasny. 2007. From risk to resilience: What role for community greening and civic ecology in cities? In *Social learning towards a sustainable world,* ed. A. Wals, 149–64. Wageningen, The Netherlands: Wageningen Academic.

Tomkins, S. and A. Collins. 2005. *Promoting optimal self-care: consultation techniques that improve quality of life for patients and clinicians* Dorset, UK: Dorset and Somerset Strategic Health Authority.

Tones, K. 2001. Health promotion: The empowerment imperative. In *Health promotion: Professional perspectives,* ed. A. Scriven and J. Orme, 3–16. Basingstoke, UK: Palgrave.

Tones, K. and J. Green. 2004. *Health promotion: Planning and strategies.* London: Sage Publications.

Transition Town Health and Wellbeing Group. http://totnes.transitionnetwork.org/healthandwellbeing/home (accessed June 26, 2009).

Transition Town Network. http://transitiontowns.org/TransitionNetwork/TransitionNetwork (accessed June 26, 2009).

Transition Town Totnes HIA. http://totnes.transitionnetwork.org/healthandwellbeing/HIAintro (accessed June 26, 2009).

Wills, J. T. 2007. The role of the nurse in promoting health. In *Promoting health: Vital notes for nurses,* ed. J. Wills, 1–10. Oxford: Blackwell.

World Health Organisation. 1978. Declaration of Alma-Ata International Conference on Primary Health Care, September 6–12, 1978, Alma-Ata, USSR.

———. 1986. Ottawa Charter for Health Promotion. Paper presented at First International Conference on Health Promotion, November 21, 1986, Ottawa.

———. 1998. *Health for all in the 21st century* (A5/5). Geneva: WHO.

———. 2008. *Climate change and health* (EB 122/4). Geneva: WHO. http://www.who.int/globalchange/climate/en/ (accessed June 26, 2009).

———. *Climate and health factsheet.* http://www.who.int/mediacentre/factsheets/fs266/en/index.html (accessed February 15, 2009).

12 Weaving Change
Improvising Global Wisdom in Interesting and Dangerous Times

Wendy Agnew

LABYRINTH

> The spider's touch, how exquisitely fine!
> Feels at each thread and lives along the line.
>
> (Pope 1733)

I am at the computer, on the web at the edge of the Luther Marsh where I live. I'm a spider sticky with curiosity, stalking the healing potential of eco-education in planetary crisis. Will my intention help convert these times *to* 'interesting' *from* dangerous? The web shivers; an idea lands and a spot of praxis moves me out of my dream. I lunge forward, passionate for survival. The computer divulges this formula . . .
Passion dot Global dot Com—Passion dot Global dot Com ~ Passion~~

Figure 12.1 Teacher as Proto-Shaman.

What pattern is this? A twisting trope? I morph from spider to Ariadne of Greek legend clutching a ball of red wool; my plurivocal umbilicus. It stains my fingers. A labyrinth yawns white under a seething sun—Tugging gently on the fecundity of solar plexus, I dye the cord black, cut it into surfaces, lines, and words . . .

Ariadne is a transformative archetype. She floats in the ocean of mythology, the placenta of consciousness. Her grandfather is Helios—the Sun titan. Her gift of yarn let Theseus, the killer of the fabled Minotaur, thread his way through a deadly labyrinth. She is a goddess of fertility, snakes, and rebirth and one who weaves us through confusion (Graves 1976, 329)—will the labyrinth of climate change challenge her? . . .

THRESHOLD

> Daedalus in Cnossos once contrived
> A dancing floor for fair-haired Ariadne.
>
> (Homer in Graves 1976, 329)

I dance into this chapter with the distaff of a question and a blank. Without the space for relationship I am just spinning wheels. My premise for ecotransformative education in interesting times is that warp of personal passion supports weft of com-passion. So I ask the reader to take a modest moment, and insert the thread of a word upon this fragment, a word that resonates to heart, to soul, or to the core of your being at this very instant, a word as simple as "coffee." If it is, you should probably go and get one (fair trade, non-GM) before we start weaving.

~ ———————————————————— ~

Thank you . . .

Past (word of the moment):

My educational history started with the told and mould method. But that didn't work . . . because I didn't fit, so I opted for grind and grade—blunder into blades of expectation with a tissue. Painful, but it got me through and left me with a sense of compassion for slaughtered forests. Later, I was attracted to more curvilinear learning paradigms, especially ones that suggested we follow the child, follow the planet, follow the bliss (Campbell 2004).

I'm not sure what your educational history was. My mother loved school, as an escape from a crumbling family. Her word of choice would perhaps be *archaeology*. My first boyfriend made a lot of money from school in nefarious activities that he will now deny. His word might be *botanical*.

Ariadne takes out a crumpled photograph and smiles nostalgically. It's her boyfriend Theseus. The educational past is filled with Theseii. (Is that the plural of Theseus?)—Heroic rationalists with fabulous chins and firm intentions. Ariadne's memory takes her back . . .

"Have this ball of red wool," she suggests winsomely, batting her eyelashes. Theseus gives her an A, goes into the labyrinth and hacks 'sensual depth' (the Minotaur) to pieces with his fists. He abandons Ariadne on the island of Naxos (or is it Praxis?) and she goes off with 'sensual excess' (Dionysus).

"What is this impressionistic tangle?" readers may ask.

My impressions implicate cognition in the cause and cure of climate change and suggest that healing synthesis is an improvisation of many threads.

This chapter is a dimple on the face of global wisdom, and as such, it is non-linear, narrative, complex, roundish . . .

I don't want to inundate you with analysis, especially when I'm still trying to imagine the texture of your word, so let's veer into practice as a way to sensualize this interaction. Without sensual curves of experience, the weave is flaccid, meaning unravels . . .

We have started with a blank, and speculative umbilici that will, hopefully, lead us out to the edge of the planetary loom, and back again through relational patterns of our selves; perhaps into a tapestry of social action, personal potency, and a diversity of ecommunions that will help mitigate the effects of climate change and swell the sea-change of holistic (Miller 2005) educational paradigms.

"Dr. Agnew," say the children. "Can we go to the forest?" The muscles and mind of a greater-than-human world call to them. Their word is consistently *Nature* . . .

THRILL OF TWIST AND TURN

> I have known the inexorable sadness of pencils,
> Neat in their boxes, dolour of pad and paper-weight,
> All the misery of manilla folders and mucilage,
> Desolation in immaculate public places.
>
> (Roethke 1948, 29)

Threads of desire conspire to 'curriculate.' Ariadne weaves me into cleft of verb. I am no longer noun of teacher—certain future a dissolving iceberg. I must move. . .

"Don't leave me, Theseus," I wave from the evaporating shoreline of knowledge transmission.

Above, grandfather Helios, the overexposed sun, peers down into a classroom where the fluorescent lights are intentionally turned off.

Radiant grandpa mutters, "Look to the children."

Pattern #1—I'm teaching body systems research—or in Montessori parlance, 'following the students' (Montessori 1965, 271; 1973, 171; 1998, 85), who are learning about the mysteries of corpus Sapien Sapien. The reproductive system is the hottest category. But everyone is coming onside with trips to the library and plans for animating their presentations with giant games, models, murals, field trips, placemats!? This is a blend of David Bohm's "proprioceptive thought"—thought radically oriented to situation and participants (Bohm 2003, 381); Illich's "deschooling" paradigm (1970, 1–24); Reggio Emila's "design, documentation and discourse" (Edwards, Gandini, and Forman 1998, 240); of methodology through a Montessori (child-directed) lens. This is also a nod to Heisenberg's Uncertainty Principle (1927)—the more precisely the position is determined, the less precisely the momentum is known at this instant, and vice versa. Far be it from me to stifle dynamism by inflicting precision on our cognitive collective (Zohar 1990, 29)! Surprises flow up, knowledge flows down, the loom is betwixt them both . . . the wisdom between.

We are all sitting on the floor in a circle—the way we do all our knowledge sharing—and P puts up her hand. She and her partner have a stack of books on the respiratory system. People focus on her serious intent and she says, "Last night when I was looking at the pictures in this book, I had an idea." Her voice holds something akin to wonder and we focus.

She waits for stillness and then P drops her delightful epiphany—"Lungs are the trees." She continues, thumbing through her book for the picture. "Look, you can see the branches, the roots, the stems."

People stop interrupting and commiserating, drawn by the synthesis. Someone adds, ". . . and trees give us oxygen." After several communicatory moments we come up with:

"Lungs are the trees of our body and trees are the lungs of the planet."

A nod to Habermas (1987) and his faith in the powers of "the public sphere" to generate inspiration. We have a history with trees . . .

This modest moment of discovery infuses the rest of the class with polymorphous perversity of the philosophical kind (Schwab 1978) and we try to relate all our body systems to the larger body of the animate earth. The conversation went something like . . .

"What about the circulatory system?"

"The blood of us is the rivers of the world and the heart is the moon. It pulls the water."

"No way, the moon isn't the heart!"

"The heart is the sun behind the trees." (Ariadne squints, taking her hand off the shuttle.)

"The molten core of the earth is the heart."

"Then lava is part of the circulatory system of the earth."

"Wait, the lava is part of the skeletal system because it's rock."

"What, the skeletal system is what?"

"The tectonic plates and the mountains and valleys."

Figure 12.2 One of our protest posters lashed to the trunk: "Saving a Sacred Tree," published in *The Kitchener Record*, March 2. Photo by Steve Sugrim 2009.

"What about the digestive system?" This had stimulated much euphoria the day before because of the word "anus."

"Ah . . .," " they teetered on the brink of mayhem and then a wise voice suggested, "Marshes, marshes are the digestive system because they purify the water."

"And that's part of the circulatory system too."

"So they're all connected."

"But what's the reproductive system?" Again the mayhem.

Howard Gardner (2003) winks at us through a fog of pollution drifting up from Detroit, pleased that we're engaging in intra- and interpersonal communion and communication, patterning, "languaging" (Maturana and Varela 1987, 210), even kinestheic and artistic resonance. We make pictures and go out into the field of embodied learning to play 'capture the fact'—a game invented by the students using their collective research as a way to chase down, capture, and also liberate their classmates. We are trying to recapitulate natural systems by providing opportunities for what physicist Ilya Prigogine calls "catalytic loops"—"instabilities through repeated self-amplifying feedback to create the emergence of ever-increasing complexity at successive bifurcation points" (Prigogine in Capra 1996, 184). That's my job, to nurture the bifurcation points.

WOMB, WANT, AND WIRE

> There is a Void, outside of Existence, which if entered into
> Englobes itself & becomes a Womb.
>
> (William Blake in Erdman 1970, 143)

This seems to be the emerging focus of education in this millennium.—Synthesis over analysis.—Relation over reduction.—Collaboration over

Figure 12.3   Embodied learning—racing with extinction.

coercion. Motion over Stagnation . . . At least that's the lesson we're learning from our four-billion-year-old Blue-Green Mentor. "Nothing is separate. All things are interconnected." What is the adage? If you don't listen to a problem it will hit you in the head? We are dizzy with the necessity to prepare our children for an uncertain future, but how do we protect them from the furnace of Helios?—The wild enthusiasms of melting icecaps?—Sudden disappearances of biological and ecological diversity?

Ariadne coughs modestly and hands us a bit of yarn. We lay it on the grid of our anxiety and it forms these words:

> *Honor the characteristics of complex adaptive systems as the cognitive and practical lacuna of modern and post-modern civilization. Mind/body/psyche is dangling from the thread of this realization, and unless nature is welcomed as the progenitor of culture, we are doomed* ~~~

Complex Adaptive Systems—Nine Principles:

1. Emergence (not planned growth, but responsive and fluent)
2. Coevolutionary (parts are developmentally influenced by each other)
3. Suboptimal (competitive energy is replaced with cooperation—not having to be the best)
4. Requisite Variety (the more diversity, the stronger and more resilient the system becomes)
5. Connectivity (all aspects of the system are radically embedded in each other so what happens to one, happens to all)
6. Simple Rules (holistic logic)
7. Iterative (part revealed in whole)
8. Self-organizing (controls are inherent)
9. Nested Holarchy (subsystems embedded in systems and so on)

(Fryer 2006)

The principles are eloquent in systems of education that opt for mixed staging (children learning in three-year age increments thereby providing multiple mentoring, community continuum, participatory democracy), freedom of speech and movement creating a society of student-based learning, story-infused curriculum in which the children play the roles of the cognitive collective in embodied knowing, nature immersion (children absorb and iterate the wisdom of complex adaptive systems through intimacy with place). Key stories and lessons are offered in a prepared environment replete with natural light, aesthetic beauty, fluent boundaries, and 'materialized abstractions' so children can self-teach. It's a cognitive old-growth forest model, so far away from the industrial monoculture that its creative chaos makes me dance.

Pattern #2—My friend's class is following principles 1 through 9—student-generated research within an arts-based (Diamond 1999), ecosensitive frame

Figure 12.4   Children poetizing with three-dimensional letters, Montessori Children's House, Tehran, 2005.

flexing with "anticipatory democracy" (Toffler 1971). They've constructed a hallway presentation on culprit vehicles. Parents wander by and stop, wiping the guilt from their SUVanity on justifications of safety and convenience. The children are ruthless. They criticize bad environmental choices. The parents smile benignly and the Hummers hum. The children offer alternatives, touching a page from Janine Benyus' *Biomimicry* (2002). The heat, beat, and treat method of industrial destructivity is obsolete. Nanotechnology has the potential to create clean, efficient, and affordable conveniences for your millennia needs, but until then, stop idling, buy modest, shop local, walk it, bus it, train it, and carpool. A monkey screams in a shrinking jungle. The disconnect is gargantuan. The earth's anguish is real to the children, their structures for making change are elastic (Hebb 2002, 60–79), they run barefoot in the muck of hope advocating "voluntary simplicity" (Elgin 1981). We send a petition to our Government. Polite nods . . . One child says, "Kyoto [Accord] is kinda like, 'we'll see dear . . . but you never do' . . . Then someone buys a 'Smart Car'."

   "Thin edge of the wedge," Ariadne smiles, nudging her shuttle between threads of internal combustion. We celebrate this coevolutionary impact. Emerging from further awareness, our junior high class deepens a long-term bond with a local farm. We start to shop at the local market, supporting

Figure 12.5   Greater-than-human-gestalt.

organic growers and local transport systems creating less pollution, closer connections . . . nested holarchies bloom . . .

Ariadne nods to her kinswoman—The Lady of Shalott, who weaves the world from her obdurate tower of glass and stone. In Alfred Lord Tennyson's poem (1833), the lady expresses the polyvalent dangers of mind/body and planet/person disconnect. Cursed to interpret life only through a mirror, she can never look directly at the face of reality but peers into the enchanted silver of her fate where images float just out of reach. Industrial nations have created a similar mirror in which "world" is a reflection of inflated want. It hovers beyond wisdom, as resource only, and climate change is its dubious progeny.

The industrial stance, featuring market economy, homogenization, 'diminishing marginal return of utility,' is reinforced by standardized curriculum that supersedes local, tribal, indigenous epistemologies. But the mirror is cracking. The tower crumbles in catalytic validity with permafrost (Dyke 2007). The Lady of Shalott throws her spindle out the window. It hits Sleeping Beauty on the head.

"Take a look at the roots of meaning that extend into pre-conquest consciousness (Sorenson in Wautischer 1998, 98–105)" . . . she shouts from the edge of REM sleep . . . "Dreaming, and imaging (Pink 2005) . . . potent stuff," she adds, running towards the visual diversity of the children's trees. Ariadne digs in her pockets for the requisite "stuff of dreams." Her fingers close on the words of ecopsychologist Stephen Aizenstat:

> at the dimension of the world unconscious, the inner subjective natures of the world's Beings are experienced as dream images in the human psyche (in Roszak 1995, 96).

From the proscenium of these words comes a chorus of anthropologists, psychologists, biochemists, ecophilosophers, theologians, quantum physicists, and educators who iterate the Gaia Hypotheses (Lovelock 2006) that "earth is an animate system and consciousness is a field of radical connectivity

Figure 12.6   Writing on the cave walls of urban dystopia.

shared by the process of being" (Varela et al. 1993, 213). The Big Other of 'Globalization' lumbers along, stepping on the feet of this humility with corporate mayhem, holding the cracked mirror of affluence up to 'developing' nations. It stutters ridiculously, "Want, want, wanton, but winnow the will." Through a relational holism that takes relationship as having primacy over entity (Teller 1986, 71–81) and as a way to heal our "Affluenza" (Hamilton and Denniss 2005), we begin to research manufacturing origins, processes, and impacts. The "birthing" of "product and the toxins of bi-product."

MARGINS OF METAPHORS

> For knowledge add, for wisdom take away
>
> (Wright in Zwicky 2003, 87).

Totem, Chant, and Ritual are the anchors and ballasts of human culture (Agnew 2002, 153). To make any valid paradigm shift, these beg to be handled, honed, and honored. Historically, we have forced the conquered (including our animate planet) into docility by imposing new mantels of meaning on the totems, chants, rituals that define and sculpt culture. The Celts, my ancestors, were compelled to trade bards for bibles, Ogham for Alpha (Graves 1976, 113–22). The Hawaiians got the dubious bargain of syphilis for love. The First Nations of the North Americas were incarcerated in residential schools and given pencils for arrows, beads for land, iron for information, mannequins for masks. Ironically, the word Totem (an Ojibwa word meaning guardian/kinship) is now used as a marketing tool for extrusions of matter called 'stuff'; chant transforms to 'slogan'; ritual becomes 'habit or addiction' (Agnew 2002, 160). Father of sociology Emile Durkheim wrote, "God is society writ large" and campaigned for "organic solidarity" (1915), but as our money markets shift and our ecosystems shrink we are starting to realize that God is more than society. God is Gaia. Without her, we are but space debris.

The subjugation of nature cultures and perhaps its corollary, childhood (Montessori 1998, 72), is a history of violence beginning with the necessary Socrates (Abram 1996, 116), who helped nudge the 'reason' chorus into Descartes' constrictive mind/body dupion. The thrilling notion of mind over matter quickly (in earth terms) transmogrified into radical materialism and slogans like, 'conspicuous consumption' and 'planned obsolescence.' According to economist George Reisman, "the problem of the millennium is erroneously believed to be how to expand the desire to consume so that consumption may be adequate to production" (Reisman 2006). Sadly, schools duplicate this pattern as they pour young minds into the standardized commodity of curriculum moulds (Illich 1970). Education, as a dubious exercise of cognitive consumption, exacerbates the problem of climate change. The delicate weave of ecoculture is deeply frayed every time a buzzer rings to interrupt students from their focus, every time we close the doors on going out into nature to learn, every time we impose subject blocks of curriculum (Perlstein 2007) and forget the thematic nature of knowledge as a manifestation of the story of complex adaptive systems. Without radial identification with nature, docile bodies (Foucault 1975, 138–9) of growing minds reverberate to echoes of damaged biosphere. Change begins in the cartilage of culture that nurtures the spine of nature. Not necessarily the ideology but in the interface between thought, feeling, physicality, and environment . . . in living symbols and rhythms that have been pushed to the margins of metaphor.

> The difference between symbol and metaphor has to do with interpretation. Metaphor belongs to the cultural context in which it is elaborated, and can therefore be decoded with relative normative certainty. The symbol is mysterious, fleeting and unattainable. One of the 'catastrophic' events that can occur in a given culture is the degradation of the symbol to metaphor . . . The obsession that seeks to make everything clear is the death of poetry. (Zignani 2008)

"Symbols may be considered the poetry of matter," suggests Ariadne.

Pattern #3—We've taken the junior highs to a wilderness camp. Engaging in methodological pluralism, naturalistic enquiry (Eisner 1991, 2), and active semiotic engagement (Selby 2006, 351–65). We've done this many times since they were young, but this time we have a guide, C, who understands relationship, in its emic (bio-articulate) (Salthe 1985, 288) potency. We gather in silence, that's a prerequisite, to calm the corpus callosum (Bolte Taylor 2008). Imagine the brain and its delicate folds, convolutions, and conjunctions are tracks in the wilderness, but our twentieth century brains have tracks, not only widened and straightened by commercial expediency but the flow between hemispheres is loaded with cantankerous bureaucracy. Are we gridlocked in consumerism, as Ivan Illich suggests, prisoners of addiction and envy (Illich 1973)? We can't reform our actions unless we reform our cerebral pathways and address what Richard Louv terms "nature deficit disorder" (2005). That's what happens here. We play a game of sardines in the pitch dark. The children thrill to the sudden use of sluggish senses. The leaves are pungent, the sounds of the night envelop us in the vivid reality of Habermas' "Life World" (1987). There are no fluorescent pixels to cancel out the music of our hearts. After the game the blood is singing and the mind is alert. There is a collective intensity of noetic inclusion—that consciousness is a factor of all being.

C gathers us and throws the cloak of mystery over our shoulders. We will wait in song at the base of the hill until everyone reaches the top. We begin to sing in a language lost to us but found in primal rhythm. C touches someone and a scrap of the song detaches and begins to climb the hill. The way is lit with tiny candles. This ritual is like a birth of sorts. We, at the bottom, watch and listen as the dark shapes of our friends trickle the melody up between the lights to strengthen at the summit. We are finally met at the top of the hill where a large fire burns. The song swells. Everyone is singing (a miracle in the age of cool and self-consciousness). C begins to thank the place. She is talking to other than human entities and we are agog. There is nothing strange or funny, only surprising and right. C invokes the blessings of the four directions. She takes the pelvic pan of a large mammal, places glowing coals in its basin, holds it up and makes a simple prayer to the North, the earth. There is utter focus. There is a memory in our own bones that trembles on the brink of the ridiculous but believes in "life symbols, the roots of sacrament" (Langer 1963, 145). C looks, with clear eyes into our midst. "Who wants to go next?" She asks this simply, as if we had been doing this ritual since we were born. There is a long pause and some shifting and then M, one of the students, stands up and takes the Hawk feather. He goes to the edge of the firelight. "I invite the East, the wind into our circle," he improvises with panpsychic brevity. And as if in answer there is a rustling of leaves above us.

What happens when the fabric of self shudders and expands? In *Stroke of Genius* (2008), neuroanatomist Jill Bolte Taylor describes the "right brain nirvana" she experienced when a blood vessel ruptured in her left hemisphere:

> I was immediately captivated by the magnificence of energy around me. And because I could no longer identify the boundaries of my body, I felt enormous and expansive. I felt at one with all the energy that was, and it was beautiful there . . . the more time we spend choosing to run the deep inner peace circuitry of our right hemispheres, the more peace we will project into the world and the more peaceful our planet will be. (Bolte Taylor, 2008)

Did we, standing on that hill, awash with the connective potency of rhythms and symbols, simulate some remembered epiphany now clogged by the infinite regress of serial analysis? In contrast to the left hemisphere, the right hemisphere functions as a parallel processor. It works through synthesis, as stories do.

A COUPLE OF SNAGS

> Mustard flowers,
> no whale in sight,
> the sea darkening.
>
> (Hass in Zwicky 2003, 100)

As you can tell, I'm embroidering this chapter with scraps of narrative—tribute to geologian Thomas Berry's imprecation to expand the ecozoic story (Berry 1999, 133). Granted, this will not be the apocryphal opus we need to heal the world psyche, but as a drop in the bucket it swells the pool . . . or the thermohaline collapse. A particle of article floats poignantly on the tides of tomorrow. It reads . . .

Figure 12.7   Death: nature-based production of Ovid's *Metamorphosis*.

Human civilization began with the stabilization and warming of the Earth's climate. A colder unstable climate meant that humans could neither develop agriculture or permanent settlements. With the end of the Younger Dryas and the warming and stabilization that followed, humans could learn the rhythms of agriculture and settle in places whose climate was reliably productive. Modern civilization has never experienced weather conditions as persistently disruptive as the ones outlined in this scenario. As a result, the implications for national security outlined in this report are only hypothetical. (Schwartz and Randall 2003)

As hypothetical oceans swell and plans to protect national interests intensify, Ariadne mutters, "There are snags in our thinking." Reversing the action of the treadle, she is profoundly aware of the principle of connectivity, so concerns about national security makes a tangle.

"Is this not a Global Issue?" She muses as she plucks a carbon covered thread from the twists and turns of corpora colossa, "The industrialized nations are the major culprits of climate change," she weaves as smoke-stacks belch and forests tumble. Theseus swings into the picture, geo-engineering (Schneider in McIntosh 2008, 77), helmet gleaming, brandishing his sword, shouting, "The G-8 nations have the most to lose, back all of you, back!" But he stubs his toe on the Rowland-Molina Hypothesis[1] and is sucked through the Ozone hole. Ariadne plucks that thread out of the picture and ties it around her finger. "Remember to patch the section on interconnectivity, interplanetary relationships, ecological unconscious (Roszak 1992, 32), structural violence, the difference between mentorship and mastery," she tells herself.

Pattern #4—I'm in a global labyrinth following the principles of complex adaptive systems over the horizon of arts-based, nature-framed research (Agnew 2007, 329–97). Our curriculum is fluent, like a rising tide. E has chosen to 'become' his off-the-grid grandfather for a dramatis persona. He rivets us to wind turbines, solar panels, and acres of organic garden. Instead of focusing on 'dominant' civilizations for cultural research, the upper elementary classes have chosen to explore disappearing tribes and earth wisdom through dramatizations and mask work. We simulate vision quests and begin dialogs inspired by traditional healer of the Carrier Nation, Sophie Thomas (Jacks 2000). We explore the possibility of a roof garden both as insulation and tribute to botany. T, a junior high student, proposes we use their play (an environmental adaptation of *A Midsummer Night's Dream*) as a fundraiser for 'Sea Shepherd.' We form a Mandate and rename it Earthdate: *To Act in accordance with the cycles and necessities of our global heritage and promote healing of the biosphere in every thought and deed.*

Perhaps there are no national boundaries to be protected, only econational bonds to be reconnected. Perhaps in our schools, those bonds begin with ages, classes, subjects, themes, stories, communities, countries, wilderness, and psyche.[2] Mentorship demands a listening stance.

ECO-ACTIVE LISTENING

> Consciousness walks across the land bridge on the deer's stare into the world of things. This is knowing. It tastes of sorrow and towering appetite.
>
> (Lilburn in Zwicky 2003, 76)

Docile bodies are part of the supervenient collapse—the fall away of the relationship between cognitive and emotional systems and physical

Figure 12.8   Spelling, "In the Magical Sense" (Selby 2008, 259). Researching the Celtic nether mind (McIntosh 2008, 248), we search for our veins in shadows.

systems—experienced when wilderness vanishes. Systems of conscious-ness—I emphasize the nonduality between mind and matter (Bohm 2003)—respond by recreating wilderness and wildness through instability . . . radical climate change . . . leading to radical and unexpected connec-tivity (Briggs and Peat 1989). Children recapitulate patterns of instability as evolutionary bifurcation points within their own psychophysical devel-opment (Montessori 1997, 59–70). Adolescence and early childhood are especially potent with potentials for growth, innovation, and/or regress. The future tilts upon the thread of this kinship.

"Children are the ones passionate for environmental rapprochement," Ariadne murmurs stretching her own arms. "Uppity evolutionaries," she cocks her head and listens . . .

"'Ecommunion' can massage the body/mind/earth tangle." Ariadne hooks a wayward fiber of trust upon this infant term and her fingers fly.

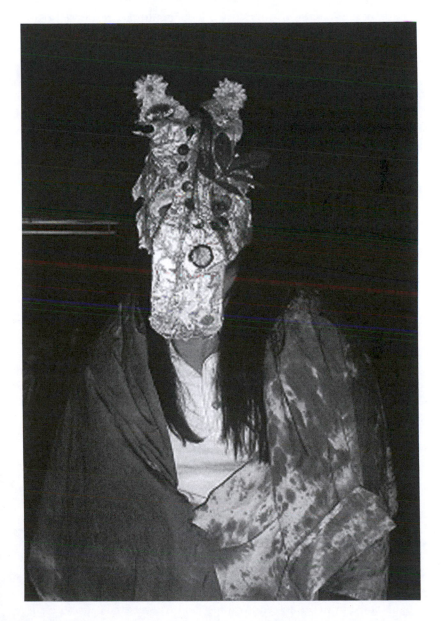

Figure 12.9   I am the Earth Horse . . . treasured God of the Celts from long ago. I am still living in the Earth as a type of nutrient for plants and animals alike . . . Excerpt by my student Myrony Auarian from her virtual vision quest, 2009.

Ecommunion nudges its breast into the breach between science and art, its belly pressed against the curve of biosphere as "phase beauty . . . reflected in every completed subordinate rhythm, and tolerant of numberless variants. . . " (Langer 1988, 86). It is a new creature, as are we . . .

As students and educators, we push through a brutal postcivilized stage into the ecoreflexivity of radical love. From a preconscious time of world as divine, to an ego-conscious time of world as inanimate resource, to a lyric time of connection predicated on world as embodied mind (Berry, 1999, 49; Varela, et al. 1993, 213).

I return to my computer screen. A real spider, tiny and fragile, confronts the worldwide web . . . Its octagonal romance twists my bilateral symmetry into *Passion dot Global dot Com ~ Passion*. We have the imagination to be cohesively passionate for the delicate orb of which we are one manifestation. Even if some of our mind is unwoven from that necessary romance. Do other cultures weave truth with more elegant knots? Curiosity flows . . .

OTHER WAYS ~ OTHER WEAVES

> Only love.
> Only the holder the flag fits into,
> and wind. No flag.

> (Rumi in Barks 1997, 29)

Pattern #5—I'm in Iran; a cradle of civilization. Perhaps a birthplace of psychic metamorphosis. We have started a little Montessori school and are travelling to Kerman to find culturally relevant materials at the bazaar. It is air-conditioned—by geometry rather than electricity. Ancient architects have used the vault and dome method to fractal heat and light into cool and deep. Ariadne breathes on a handful of plastic mud and a soft city forms (Khalili 2001). In Iran, gardens still enhance the local economy of urban farmers. Fountains spurt twenty meters into torrid air from centuries-old, spiral-formed artesian wells. Pigeon-houses, beautiful adobe formed bird-co-ops, fertilize nitrogen-poor fields. Evidence of pre-industrialized civilization with a soupçon of earth-reverence calls me. Nomads hover on the periphery of villages and weave archetypal wisdom into carpets. The cock is time, the garden is paradise, the vase is peace and tranquillity, and the tree of life is understanding and truth (Ahmadi 1997).

Ariadne whispers in fluent Farsi, "If climatic and social instability precipitated a great shift in consciousness several thousand years ago (Jaynes 1990, 168), then current transmutations of biosphere poise us for another great shift." To see and breathe the beauty of ancient wisdom makes me realize that post-modern mind can weave its way around the corners of its current mythology, from ego to eco. Ariadne's loom covers the page with curved shadows. Children are inducted into their first day through the burning of sacred plants. Our little Iranian school flourishes in a cultural interweave that honors the poetry of elements and the morphic resonance of ritual—ritual creates psychophysical impressions in the field of time (Sheldrake 1988, 371).

Vestiges of right-brain predominance are accessible to the post-modern mind through international syntheses of science, poetry, and art, and their conduit is language.

If consciousness is a transformative and transforming thread of our bio-cultural continuum (Cobb 1977, 53), then educating for climate change must predicate on earth wisdom, a love that reconnects the mythos of psyche to prima mater.

> To defend poetry means to defend a fundamental gift of human nature, that is, our capacity . . . to experience astonishment and to stop still in that astonishment for an extended moment or two. (Sagajewski in Zwicky 2003, 108)

I was invited to the theater. The play was about the first recorded poetess of Persia, Rabia Balkhi. She fell in love with her brother's musician, a slave, and they were both condemned to death. The night before her execution, she slashed her wrists and wrote her love with blood on the walls of her cell. "The circulatory system is the rivers, Dr. Agnew. . . , " the children murmur from yesterday. We go to a garden in Tehran and do silent meditations to nature and then create poems. This lifts a burden of sadness and prepares us to act. As a body of educators, we move into a numinous state (McIntosh 2008, 241). This is, for me, cultural and global psychotherapy. Trading poetry is a hammock over the chasm of global dystopia. When educators become voices of earth, the children find ecommunion. Parastoo, with her gift of words, stitches her dreams to the picture of our dinosaur mural *Racing with Extinction*. She meta-muralizes. The weave deepens. . . .

For the Mural Dinosaur

Once the Mural Dinosaur
Pokes its head through the clouds
A child within it sings a song
For tropical forests,
The tiny pineapples,
Dusty blistered feet of African child,
The call of peddler kids
And little hawker shoeblacks
Who carry huge volumes in their hearts.
A song of sleeping love
And nobody knows
the mystery of waking it up
But the Mural Dinosaur

–By Parastoo Asef—Tehran/August, 2008

Figure 12.10   Ecommunion—In the garden.

If you are reading this book, you are probably a convert to eco-education as necessary ballast to an uncertain and dubious planetary future. So, instead of convincing you of the necessity for impassioned intervention in an avalanche of human folly, I will offer you the completion of a story and a vision of the future. Ariadne shrinks back into her spider persona, delicate on the compass of a lurching world. She is tiny, but her 'i' is ubiquitous in the 'we' of weave. She says with the last breath of her humanness . . . *"Com—pass—i—on"*.

CONCLUSION

> Every act in language brings forth a world created with others in the
> act of coexistence . . . thus every human act has an ethical meaning
> because it is an act of constitution of the human world . . . we have
> only the world that we bring forth with others, and only love helps
> us bring it forth.
>
> (Maturana and Varela 1987, 210)

Pattern #1—Continued . . . From the fecund chaos of our discussion L
shouts, "Muscles, Muscles is gravity pushing down and up on the earth!"

I feel that pressure.—The gravity of the situation. A soft voice intrudes
upon my angst. It's L, diagnosed several years ago with a serious learning
disability, ". . . and the nervous system," he says quietly, ". . . the nervous
system is the myths. . . the stories . . . "

It's the end of the class and everyone gathers their books in preparation
to leave. The portable is nestled in snow so the trip back to the school is a
glorious adventure. There is one little girl sitting still in the flurry of activity. She looks into my eyes and reminds me what she said last week. The
students pause. The air is open. There is a blank and A fills it with this:
"The earth is a cell, Dr. Agnew. The earth is a cell in the body of the Universe." And she holds up a picture of the blue-green planet redolent with
cloudy potential.

SUN AND MOON SCENARIOS

> . . . not to our easy instincts, but to our better angels . . .
>
> (President-elect Barack Obama,
> Whistle Stop Rally, Baltimore, January 17, 2009)

Sun Scenario

I'm writing the final part of this chapter in a record thaw. The crows are
back too early. It's just after Christmas, Winter Solstice . . . The Festival of
Light, and there is no power at the Luther Marsh. The Sun is light and heat
for us, and the woodstove is warmth. A flashlight cuddles up to my pencil
stub—computer is a non-option. The community gathers closer in such
times. Neighbors sharing dinners, generators are borrowed so the vet can
see to stitch up a cut horse. Will our manufactured disasters bring us closer
to a generative postcivilized wisdom? I close my eyes:

It's 2050. Power outages are common now, but our school has turned to
wind and solar—a project designed by students and funded by a governmen-
tal/parent group. In true human-as-great-adaptor form we are nuanced by
necessity. Gas has risen again and public transport is an economic necessity.

Trains, once bought out by oil companies (Black 2006), proliferate. The globe shrinks as it melts, dissolving boundaries. Nations look to each other for help . . . Iran, Israel, and Canada form an exchange based on green technologies—woven and adapted from ancient innovations (WebEcoist 2009).

Students work on survival projects iterating my dad's youth during the Great Depression—firefighting, planting, community action gardens, recycling dirty technologies to green (Ecomotion 2009). Instability mothers invention as "school" (from schola—L. leisure), becomes a bifurcation point for cognitive village and genesis of micro–economy. Interdependent embryonic communities bloom. Students are steeped in the identity of service. Consciousness shifts in eco-socio-motion.

Education has evolved into an autopoietic (Maturana and Varela 1974) system—where poetry (or dynamic connectivity in a greater-than-linguistic sense) is systemically enfolded in a self-manifesting dialectic of form and function (such as a living cell, an ecosystem, a sustainable community). Allopoietic systems (the cause of climate change and previous model for industry and mass-compulsory schooling) are outlawed. All systems where function subdues, degrades, or pollutes form (as in assembly lines, strip mines, clearcutting, rampant development to the detriment of biosphere) are classified as crimes against the future. These habits have died hard.

We, in our ELS (emergency learning shelter), learn mainly through experience—building a solar generator so neighbours can have heat and light, food, and the odd movie. We sleep at the shelter often. My mother whispers of the war and how close people came in disaster . . . The Sun— the big Aster (star)—Helios is hotter now. I'm lying in a sleeping bag looking up at a mobile of the solar system held together by red twine. Helios winks and murmurs, "I've seen it all before." Ariadne is back. She tugs me down to the labyrinth of sleep and dream.

Moon Scenario

Before the Arabs invaded Iran, Zoroastrianism was the dominant faith. You can see the winged Sun symbols carved into the columns at Persepolis. An Iranian friend explained, "They didn't worship fire, but the essence of light." The moon reflects some of my fear through the light of future. Part of the essence of light is the knowledge of darkness. When the globe shrinks and the boundaries contract there will be skirmishes—razor wires glinting on the identity of night fires. Sharp threads tear into the fabric of knowledge. Raiding parties spill into libraries burning books for heat. The kids have guns. The titans eat their children . . . A new paranoia is born of scarcity. Learning is prescriptive and conscriptive. Schools work to protect the interests of giant corporations by cognitive and military indoctrination. Resources are caged. Thought is interrogated. Movement is restricted. The earth as a living and articulate system is silenced in the unrelenting conspiracy of consumption . . . but climate change continues.

I wake up and look at the solar system mobile—glad to be part of the solution. Someone is passing out blankets made of red wool and singing an old Celtic tune:

> If it wasna for the weavers what would ye do?
> You wouldna hae a claith that was made o' woo'
> You wouldna hae a coat o' the black or the blue
> If it wasna for the work of the weavers.

(Blood and Patterson 1992, 261)

NOTES

1. At the time (1974) a much disputed theory implicating human activity in ozone depletion.
2. The templates are there: vertical age grouping, qualitative and peer assessment, nature-based learning, story-inspired curriculum. One of the best examples is "The Montessori Hershey Farm School" in Burton, Ohio, Ewert-Krocker, "The Hershey Montessori Farm School," *The NAMTA Journal* 26 (3) (Summer 2001) 377–453.

REFERENCES

Abram, D. 1996. *The spell of the sensuous: Perception and language in a more-than-human world.* New York: Vintage Books.

Agnew, W. 2007. *The universe is a horse: Autopoietic education for technoprosaic times,* University of Toronto.

Agnew, W. 2002. *A chronicle of stumbles.* University of Toronto.

Ahmadi, R. T. 1997. *Symbolism in Persian rugs 1997.* http://www.islamicmanu-scripts.info/reference/articles/Ahmadi-1997-MO-03-1-Symbolism.pdf (accessed December 12, 2008).

Benyus, J. 2002. *Biomimicry: Innovations inspired by nature.* New York: Harper Perennial.

Berry, T. 1999. *The great work: A way into the future.* New York: Bell Tower.

Black, E. 2006. *Internal combustion: How corporations and governments addicted the world to oil and derailed the alternatives.* New York: St. Martin's Press.

Blake, W. 1970. *The poetry and prose of William Blake.* David Erdman ed. Garden City, NY: Doubleday.

Blood P. and A. Patterson, eds. 1992. *Rise up Singin.'* Bethlehem, PA: Sing Out Corp.

Bohm, D. 2003. *The essential David Bohm.* Lee Nichol ed. London: Routledge.

Bolte Taylor, J. 2008. *My stroke of insight: A brain scientist's personal journey.* http://www.wanderings.net/notebook/Main/StrokeOfInsightJillBolteTay-lorOnTED (accessed January 23, 2009).

Briggs, J. and F. David Peat. 1989. *Turbulent mirror: An illustrated guide to chaos theory and the science of wholeness.* New York: Harper & Row.

Campbell, J. 2004. *Pathway to bliss: Mythology and personal transformation.* Novato, CA: New World Library.

Capra, F. 1996. *The web of life: A new scientific understanding of living systems.* New York: Doubleday.

Cobb, E. 1977. *The ecology of imagination in childhood.* Dallas: Spring Publications.

Diamond, P. and C. Mullen. eds. 1999. *The postmodern educator: Arts-based inquiry in teacher development.* New York: Peter Lang.

Durkheim, E. 1915. *The elementary forms of the religious life.* New York: Free Press.

Dyke, L. D. 2007. Landslides and stability of permafrost slopes in a warming climate. *Natural Resources Canada.* http://gsc.nrcan.gc.ca/permafrost/suppdoc_e. php (accessed January 12, 2009).

Ecomotion. 2009. *An ethical networking diary connecting students, communities, charities, green businesses, and those committed to change.* http://www.ecomotion.org.uk/ (accessed February 5, 2009).

Edwards, C., L. Gandini, and G. Foreman. 1998. *The hundred languages of children: The Reggio Emilia approach—Advanced reflections.* London: Ablex.

Eisner, E. 1991. *The enlightened eye: Qualitative inquiry and the enhancement of educational practice.* Toronto: Collier Macmillan.

Elgin, D. 1981. *Voluntary simplicity: Toward a way of life that is outwardly simple, inwardly rich.* New York: William Morrow.

Ewert-Krocker, L. 2001. The Hershey Montessori Farm School. *The NAMTA Journal* 26 (3) (Summer) 377–453.

Foucault, M. 1975. *Discipline and punish: The birth of the prison.* Paris: Gallimard.

Fryer, P. 2006. *What are Complex Adaptive Systems?* (cited December 14, 2008). http://www.trojanmice.com/articles/complexadaptivesystems.htm.

Gardner, H. 2003. Multiple intelligences after twenty years. Paper presented for the American Educational Research Association, Chicago, IL, April 21.

Graves, R. 1976. *The white goddess: A historical grammar of poetic myth.* New York: Octagon.

Habermas, J. 1987. *The theory of communicative action, Vol. 2, Lifeworld and system: A critique of functionalist reason.* Boston: Beacon Press.

Hamilton, C. and R. Denniss. 2005. *Affluenza: When too much is never enough.* Sydney: Allen & Unwin.

Hebb, D. 2002. *The organization of behavior: A neuropsychological theory.* Mahwah, NJ: Lawrence Erlbaum.

Heisenberg, W. 1927. "Über den anschaulichen Inhalt der quantentheoretischen Kinematik und Mechanik", *Zeitschrift für Physik,* 43, 172–98.

Illich, I. 1970. *Deschooling Society.* New York: Harper & Row.

Illich, I. 1973. *Tools for Conviviality.* New York: Harper & Row.

Jacks, T. 2000. *The Warmth of Love: The Four Seasons of Sophie Thomas.* VCR. Vanderhoof, BC: Sophie Thomas Foundation.

Jaynes, J. 1990. *The origin of consciousness in the breakdown of the bicameral mind.* Boston: Houghton Mifflin.

Khalili, N. 2001. *Racing alone: Fire and Earth, A visionary architect's passionate quest.* Hesperia, CA: Cal-Earth Press.

Langer, S. 1988. *Mind: An essay on human feeling.* Baltimore: Johns Hopkins.

Langer, S. 1963. *Philosophy in a new key.* Cambridge, MA: Harvard University Press.

Louv, R. 2005. *Last child in the woods: Saving our children from nature-deficit disorder.* New York: Algonquin Books.

Lovelock, J. 2006. *The revenge of Gaia: Why the Earth is fighting back—and how we can still save humanity.* Santa Barbara: Allan Lane.

Maturana, H. and F. Varela. 1987. *The tree of knowledge: The biological roots of human understanding.* Boston: Shambhala.

McIntosh, A. 2008. *Hell and high water: Climate change, hope and the human condition.* Edinburgh: Birlinn.

Miller, J. ed. 2005. *Holistic learning and spirituality in education: Breaking new ground.* Albany: State University of New York Press.

Montessori, M. 1973. *The absorbent mind*. Madras India: Kalakshetra

Montessori, M. 1998. *The discovery of the child*. Oxford: Clio Priess.

Montessori, M. 1997. *From childhood to adolescence*. Oxford: Clio Press.

Montessori, M. 1965. *Spontaneous activity in education*. New York: Schocken Books.

Perlstein, L. 2007. *Tested: One American school struggles to make the grade*. New York: Henry Holt & Co.

Pink, D. 2005. *A whole new mind: Moving from the information age to the conceptual age*. New York: Riverhead Books.

Pope, A. 1733. *An Essay on Man, Epistle 1*, lines 217–18. http://rpo.library.utoronto.ca/poem/1637.html (accessed April 14, 2009).

Riesman, G. 3/17/2006. *Production versus consumption*. Ludwig von Mises Institute. http://mises.org/story/2079 (accessed January 4, 2009).

Roethke, T. 1948. *The lost son and other poems*. Garden City, NY: Doubleday.

Roszak, T. 1992. *The voice of the earth*. New York: Simon & Schuster.

Roszak, T., M. Gomes, and A. Kanner, eds. 1995. *Ecopsychology*. San Francisco: Sierra Club Books.

Rumi. 1997. Trans. Coleman Barks. *The essential Rumi*. New Jersey: Castle Books.

Salthe, S. 1985. *Evolving hierarchical systems*. New York: Columbia University Press.

Schwartz P. and D. Randall. Oct. 2003. An Abrupt Climate Change Scenario and Its Implications for United States National Security. *GBN Global Business*. http://www.edf.org/documents/3566_AbruptClimateChange.pdf (accessed December 20, 2008).

Selby, D. 2008. The need for climate change in education. In *Green frontiers: Environmental educators dancing away from mechanism*, ed. J. Gray-Donald and D. Selby, 252–62. Rotterdam: Sense Publishers.

Selby, D. 2006. The firm and shaky ground of education for sustainable development. *Journal of Geography in Higher Education* 30 (2) 351–65.

Sheldrake, R. 1988. *Presence of the past*. New York: Vintage.

Sorenson, R. 1998. Pre-conquest consciousness. In *Tribal epistemologies*, ed. H. Wautischer, 79–115. Aldershot, UK: Ashgate Publishing.

Teller, P. 1986. Relational holism and quantum mechanics. *British Journal for the Philosophy of Sciences* 37, 71–81.

Tennyson, A. 1833. The Lady of Shalott. http://www.online-literature.com/tennyson/720/ (accessed April 16, 2009).

Toffler, A. 1970. *Future shock*. New York: Random House.

Varela, F. J., E. T. Thompson, and E. Rosch. 1993. *The embodied mind: Cognitive science and human experience*. Cambridge, MA: MIT Press.

Varela, F., H. Maturana, and R. Uribe. 1974. Autopoiesis: The organization of living systems, its characterization and a model. *Biosystems* (5) 187–96.

WebEcoist: Seven ancient wonders of green design and technology. Sustarinable living, green design and environmental oddities. http://webecoist.com/2009/01/25/ancient-green-architecture-alternative-energy-design/ (accessed February 1, 2009).

Westbury I. and N. Wilkof, eds. 1978. *Joseph Schwab, science, curriculum and liberal education*, Chicago: The University of Chicago Press.

Zignani, A. 2008. Rhetorical devices in literary languages. Trans. Robert Burchill. http://courses.logos.it/pls/dictionary/linguistic_resources.cap_let_2_2?lang=en (accessed January 26, 2009).

Zohar, D. 1990. *The quantum self*. New York: William Morrow.

Zwicky, J. 2003. *Wisdom and metaphor*. Kentville, NS: Gaspereau Press.

Climate Change Education
A Critical Agenda for Interesting Times

Fumiyo Kagawa and David Selby

Resistance to new ideas increases as to the square of their importance.
—Bertrand Russell

In these last pages we attempt to capture key recurring themes and proposals of this book in the form of an agenda for climate change education. We fully appreciate that the agenda amounts to a radical departure from what currently passes for climate change education in most learning spaces, and for that reason is likely to be rebuffed and resisted, but interesting times demand that we contemplate and, yes, urgently take interesting leaps in the dark. In the face of runaway climate change nothing short of a lived paradigm shift is needed.

- Education can only help allay a threatening condition by addressing root causes. The focus on presenting symptoms of carbon release and buildup that characterizes current climate change discourse and education, and the concomitant fixation on ameliorative technologies that more or less leave business as usual, needs to shift to an unpacking and critiquing of the role that the currently hegemonic economic model and social order, allied to voracious consumerism in economically wealthy societies, has played and continues to play in putting the world at risk. In addressing root causes the present tendency to see remedy as lying primarily within individual behavioral change rather than in structures, systems, and predominant worldview is likely to fall away.
- Climate change education needs to happen within interdisciplinary and multidisciplinary frames. Scientists brought climate change to our attention and we continue to devour their latest research findings but education needs to fulfill, first, a bridging role by enabling scientific intelligence to be widely and subtly understood from multiple perspectives and, second, an interdisciplinary role by helping learners at one and the same time apply cultural, social, economic, ethical, political, and spiritual intelligence to understandings of causes, implications, and proposed ways forward. Interdisciplinary approaches will also enable a reclaiming of nonscientific, indigenous knowledge.
- There can be no ethical and adequately responsive climate change education without global climate justice education. Missing from

most climate change education, especially in wealthier societies, is an appreciation that the metaphorical North of the planet is primarily responsible for carbon buildup but that its effects are coming thickest and fastest to the peoples and societies of the South. The North's enclosure of the 'atmospheric commons' and position as 'atmospheric debtor' are phenomena that are less than sufficiently exposed given Northern domination of the global dialog on climate change. Education has a role in challenging and rolling back climate change injustice.

- The educational response to climate change needs to be both local and global. While there is an overriding need for learners to explore local ways to mitigate and live with climate change, to understand how to live lightly and locally, to engage in local community development, and to reflect on what is precious in local nature and culture that may well be lost with the onset of global warming, education also needs to help animate, and so become informed by, a worldwide participatory dialog that elicits and weaves together the best of indigenous, mundane, and scientific insights about what we value and how best to live according to what we value. Common wisdom allied to the imperatives of survival are likely to ensure that the new paradigm of living and learning emerging from the dialog will embrace as key values earth-connectedness, earth concern, social justice and inclusiveness, peace and human rights, health and wellbeing. Localism in response to climate change cannot be allowed to be a question of raising the drawbridge on others less fortunate.

- Wherever it takes place, climate change education needs to be a social and holistic learning process; climate change is too urgent and important to suffer 'death by formal curriculum.' Looming rampant climate change calls for flexible learning and emergent curriculum approaches that embed climate change learning and action within community contexts. The threat is too urgent to be left to cloistered school-age education, so 'all-age learning' linked to local arenas and channels of participatory democracy and directed towards effecting responsive change locally is necessary. The threat is also too urgent to any longer continue with epistemologically under-dimensioned learning confined to rational, linear, analytical, classificatory, and mechanistic ways of knowing and seeking to effect change. Employed exclusively, even predominantly, such ways of knowing are tantamount to applying disease as remedy. There is a need for the complementary and recursive use of artistic, embodied, experiential, symbolic, spiritual, and relational learning, especially in the vital educational task of reconnecting learners to the earth while enabling them to discover their (connected) identity and realize their full potentials.

- There is need for educators to urgently and radically think through the implications of the invisibility and uncertainty of climate change.

Climate change is a phenomenon spread across space and time. It is hard to know what we face from past carbon buildup. It is hard to know what effects we, as present generation, will have, and when, and where the impact will fall. Scientists tell us it is a most threatening issue but it is a threat that can seem abstract and intangible and, so, ripe for avoidance and denial mechanisms to come into play. The difficulty of envisioning a desirable future goes hand in glove with the difficulty of understanding, and unwillingness to confront, our present reality. We thus face an issue threatening the world as we know it but to which education, with its preference for handling the tangible and aspiring to the certain, seems especially ill equipped to respond. As a fundamental contribution to climate change, it seems that educational spaces should build a culture of learning awash with uncertainty and in which uncertainty provokes transformative yet precautionary commitment rather than paralysis.

Contributors

Wendy Agnew is a Montessori guide, international teacher educator, and performance artist dedicated to the necessity of anomaly. Her publications form an eclectic blend of poetry, fiction, and academic prose detailing her work on four continents. Wendy's doctoral thesis, *The Universe is a Horse: Autopoietic Education for Technoprosaic Times* (see reference, p. 237), explores both quantum and planetary implications of letting children lead in the dance of education. She follows the premise that evolution and survival involve a layering of humility, passion, and surprises found only in the cognitive pockets of the young and the thrilling pockets of wilderness. She lives in Ontario, Canada.

Virginia Floresca Cawagas is an Associate Professor in the Department of Gender and Peace Education of the U.N. mandated University for Peace in Costa Rica. Prior to this, she was Adjunct Associate Professor of the School of Education and Professional Studies, Griffith University, Brisbane, Queensland, Australia, and of the School of Educational Policy Studies, University of Alberta, Canada; also a Senior Fellow and Visiting Professor at the University for Peace, Costa Rica. Since 2003, she has coordinated several projects for interfaith dialog and interfaith education under the auspices of the Griffith University Multi-Faith Centre. An author of articles in academic journals, book chapters, and textbooks in peace education, and civics and culture, she has conducted courses and workshops in global/peace education, human rights education, and multicultural education in various countries including the Philippines, Australia, Canada, Fiji, Jamaica, Japan, South Korea, Uganda, and the United States. She holds a Masters in Educational Management from De La Salle University and a Doctorate in Education from Notre Dame University, where she was also involved in establishing the first graduate program in peace and development education in the Philippines. She has been Editor of the *International Journal of Curriculum and Instruction* and an Executive Board member of the World Council for Curriculum and Instruction.

Darlene E. Clover is an Associate Professor in Leadership Studies, Faculty of Education, University of Victoria, British Columbia, Canada. Her research and teaching areas include community leadership, adult education, feminist pedagogy/activism, environmental adult education, and social movement and community arts-based learning. Her most recent publication is *The Arts and Social Justice: Recrafting Adult Education and Community Cultural Leadership* (NIACE, UK, 2007).

Ian Davies is Professor in Educational Studies at the University of York, United Kingdom. Recent publications include the four-volume *Citizenship Education* (2008), published by Sage and coedited with James Arthur, and the Sage International *Handbook of Education for Citizenship and Democracy* (2008), coedited with James Arthur and Carole Hahn. He has extensive international experience.

George J. Sefa Dei was Chair of the Department of Sociology and Equity Studies, Ontario Institute for Studies in Education of the University of Toronto (OISE/UT), Canada from June 2002 to June 2007. For the 2007–08 academic year he was a Visiting Professor at the Centre for School and Community Science and Technology Studies (SACOST), University of Education, Winneba, Ghana. Between 1996 and 2000 he served as the first Director of the Centre for Integrative Anti-Racism Studies at OISE/UT. His teaching and research interests are in the areas of antiracism, minority schooling, international development, and anti-colonial thought. His published books include: *Anti-Racism Education: Theory and Practice*; *Hardships and Survival in Rural West Africa*; *Schooling and Education in Africa: The Case of Ghana*; *Critical Issues in Anti-Racist Research Methodologies*; *Anti-Colonialism and Education: The Politics of Resistance*; *African Education and Globalization: Critical Perspectives*; and *Schooling and Difference in Africa: Democratic Challenges in Contemporary Context*. His most recent books are: *Racists Beware: Uncovering Racial Politics in Contemporary Society* and *'Crash' Politics and Anti-Racism: Interrogating Liberal Race Discourse*. George Dei is the recipient of the Race, Gender, and Class Project Academic Award, 2002. He also received the African-Canadian Outstanding Achievement in Education accolade from *Pride* Magazine, Toronto, in 2003, and the City of Toronto's William P. Hubbard Award for Race Relations in 2003. In October 2005 he received the ANKH Ann Ramsey Award for Intellectual Initiative and Academic Action at the Annual International Conference of the Association of Nubian Kemetic Heritage, Philadelphia, PA. He is also recipient of the 2006 Planet Africa Renaissance Award for his professional achievements in the field of African education, antiracism, and youth. He also won the 2007 Canadian Alliance of Black Educators Award for Excellence in Education and Community Development. In June 2007, he was installed as a traditional chief in Ghana, specifically, as the Adumakwaahene of the town of Asokore, near Koforidua in the New Juaben Traditional Area of Ghana.

Edgar González-Gaudiano is a Senioe Research Fellow in the Institute of Social Sciences of the Autonomous University of Nuevo Leon, Mexico. A member of the Commission of Education and Communication of IUCN and Regional President for Mesoamerica (2001–06), he has been member of Board of Directors of the North American Association for Environmental Education (NAAEE). He also worked with WWF and UNESCO in projects related to national strategies for environmental education. He has written nine books and more than 100 articles in several specialized journals in the field of environmental education. He is the Latin American liaison person and a member of the UNESCO Group of Reference for the UN Decade of Education for Sustainable Development. He is also the former president of the National Academy of Environmental Education, a NGO which seeks to foster professionalization and research in this field, and a member of the Mexican Academy of Sciences. He is the main editor of the Iberoamerican journal, *Topicos en Educacion Ambiental.*

Magnus Haavelsrud is a Professor of Education at the Norwegian University of Science and Technology in Trondheim, Norway. His work deals with the critique of the reproductive role of education and the possibilities for transcendence of this reproduction in light of the traditions of educational sociology and peace research. He took part in the creation of the Peace Education Commission of the International Peace Research Association at the beginning of the 1970s and served as the Commission's 2nd Executive Secretary 1975–79. He was the Program Chair for the World Conference on Education in 1974 and edited the proceedings from this conference entitled *Education for Peace: Reflection and Action.* He served as the Carl-von-Ossietzky Guest Professor of the German Council for Peace and Conflict Research. His publications include: *Education in Development* (1996), *Perspektiv i utdanningssosiologi (Perspectives in the Sociology of Education*; 1997, 2nd edition), *Education Within the Archipelago of Peace Research 1945–1964* (coauthored with Mario Borrelli, 1993), *Disarming: Discourse on Violence and Peace* (editor, 1993), and *Approaching Disarmament Education* (editor, 1981). For more recent activities (especially in Latin America) see his homepage http://www.svt.ntnu.no/ped/Magnus.Haavelsrud/. Email: Magnus.Haavelsrud@svt.ntnu.no.

Budd L. Hall, currently the Director of the Office of Community-Based Research, University of Victoria, British Columbia, Canada, has been working in and writing about learning and social movements for nearly forty years. He is former Dean of Education at the University of Victoria, former Chair of Adult Education at the Ontario Institute for Studies in Education of the University of Toronto and former Secretary-General of the International Council for Adult Education. He has published five articles over the past three years on learning and social movements and is the author of Higher Education, Human and Social Development in North America in the *World Report on Higher Education* edited by Peter Taylor and published by the Global University Networks for Inovation in Barcelona. He is also a poet.

Fumiyo Kagawa is Research Team Coordinator at the Centre for Sustainable Futures, University of Plymouth, United Kingdom. The title of her recently-completed doctoral thesis is: *Navigating Holistic and Sustainable Learning: Challenges and Opportunities in Ongoing and Creeping Emergencies.* She completed her Masters of Education at the Ontario Institute for Studies in Education of the University of Toronto with a focus on global and peace education. Her previous positions include: Research Assistant at the Centre for Sustainable Futures; Research Assistant for Emergency Education, University of Plymouth; country-specific researcher (South Africa) with the University of Toronto–Ford Foundation Education for Global Citizenship Education Project; Graduate Coordinator for the UNICEF CARK (Central Asian Republics and Kazakhstan) Global Education Project. She is a partner in Sustainability Frontiers, an international actual and virtual center addressing climate change and sustainability education and ecopedagogy. Fumiyo is a Japanese citizen. Email: fumiyo.kagawa@plymouth.ac.uk.

Heila Lotz-Sisitka holds the Murray and Roberts Chair of Environmental Education and Sustainability at Rhodes University, Grahamstown, South Africa. She has contributed actively to the inclusion of environment and sustainability principles and programs in South Africa's new education system, and has led numerous national and international research and teaching projects. She is Editor of the *Southern African Journal of Environmental Education,* and has served as the Scientific Chair of the 2007 World Environmental Education Congress Scientific Committee. She currently serves on UNESCO's international reference group for the United Nations Decade on Education for Sustainable Development. Email: h.lotz@ru.ac.za.

Pablo Meira-Cartea is a tenured Lecturer in Environmental Education (EE) in the Faculty of Educational Sciences, University of Santiago de Compostela, Galicia, Spain. His current research interests focus on the theoretical, ideological, and ethical foundations of EE; on education and communication in relation to climate change; and on strategic development of EE. He has taken active part in the elaboration and evaluation of diverse territorial EE strategies in Spain. He cofounded and was the first president of the Sociedade Galega de Educación Ambiental [Galician Environmental Education Society]. He is a member of the teaching staff in charge of an interuniversity postgraduate program in EE jointly offered by nine Spanish universities. He is currently supervising diverse research projects in Spain, Portugal, Brazil, and Cape Verde. He has authored and coauthored over seventy monographs and papers.

James Pitt is Senior Research Fellow at the Department of Educational Studies at the University of York, United Kingdom, where he leads an MA in Education by Research with a focus on Education for Sustainable Development (ESD). Coming from a background in design and technology education, he has worked extensively in the United Kingdom and internationally in the area of ESD in curriculum development, teacher training, and research. He is also Visiting Professor at the Amur State University of Humanities and Pedagogy in Komsomol'sk-na-Amure, Russia.

Jane Reed is Head of the International Network for School Improvement at the London Centre for Leadership in Learning based in the Institute of Education, London University. She is a senior lecturer, consultant, and coach with an academic background in developing the connections between classroom pedagogy, professional learning, and school change and development. Her current research interest is in learning processes that free schools for social and environmental responsiveness. She has been studying in the field of ecology and systems thinking and its application to school improvement and leadership for the past seventeen years. In the 1990s she set up the first ecoliteracy project in the United Kingdom, inspired by the work of Fritjof Capra. In 2007 she led a research review in partnership with the World Wide Fund for Nature on the greening of school leadership for the National College for School Leadership (NCSL). She is currently codirecting the next phase of this research for NCSL to support the Sustainable Schools Grant scheme in partnership with Forum for the Future. Email: j.reed@ioe.ac.uk.

Janet Richardson is Professor of Health Service Research in the Faculty of Health at the University of Plymouth, United Kingdom. A registered nurse and chartered psychologist she has been researching patients' views of healthcare, and evaluating health service effectiveness since 1989. Much of this work focused on engaging staff and users in the development and commissioning of services using participatory approaches. She has made a significant contribution to the development of the evidence-base in complementary therapies including the NHS National Library for Health Complementary and Alternative Medicine Specialists Library. Janet has a particular interest in teaching and evaluating health service responsiveness to climate change, supported by a Fellowship with the Centre for Sustainable Futures at the University of Plymouth.

David Selby is an Adjunct Professor at Mount St. Vincent University, Halifax, Nova Scotia, Canada. He is also director of Sustainability Frontiers, an international actual and virtual center addressing climate change and sustainability education and ecopedagogy. He was previously Professor of Education for Sustainability and Director of the Centre for Sustainable Futures, University of Plymouth, United Kingdom, and, before that (1992–2003) Professor of Education and Director of the International Institute for Global Education at the Ontario Institute for Studies in Education of the University of Toronto, Canada. He has directed UN-funded and orchestrated curriculum development projects in some ten countries in South East Europe, the Middle East, and Central Asia and has lectured or facilitated workshops and seminars in global education, environmental education, education for sustainability, and related fields in more than thirty countries. He has (co)written or (co)edited some twenty books and over one hundred book chapters and articles, including, latterly, some pathfinding papers on climate change and education. His books include *Global Teacher, Global Learner* (1988); *Earthkind: A Teachers' Handbook on Humane Education* (1995); *In the Global Classroom, Books One and Two* (1999, 2000); and *Weaving Connections: Educating for Peace, Social and Environmental Justice* (2000). His most recent book is *Green Frontiers: Environmental Educators Dancing Away from Mechanism* (Rotterdam, Sense, 2008). He was the co-recipient of the Canadian Peace Education Award in 2003 and is a Fellow of the Royal Society of Arts. David can be contacted at Sustainability Frontiers <sf@interconnections.f9.co.uk>.

Toh Swee-Hin (S. H. Toh) was Professor and founding Director of the Griffith University Multi-Faith Centre, Brisbane, Queensland, Australia, which seeks to promote interfaith dialog towards a culture of peace at local, national, and international levels. In September 2009, he took up the position of Distinguished Professor in the Department of Peace and Conflict Studies of the U.N. mandated University for Peace in Costa Rica. Born and raised in Malaysia, he has taught in Australia and Canada in the fields of intercultural/international education and education for peace, human rights, justice, multiculturalism, and sustainability, and is a Senior Fellow of the UN University for Peace in Costa Rica. Professor Toh has been extensively involved since the 1970s in education, research, and action for a culture of peace in North and South contexts, including Australia, Canada, South Africa, Japan, Uganda, South Korea, and especially in the Philippines. In 2000, he was awarded the UNESCO Prize for Peace Education. He has also been active in various global networks or organizations for peace education, peacebuilding, and interfaith dialog, such as UNESCO, the International Institute on

Peace Education, the World Council for Curriculum and Instruction, the Asia-Pacific Centre of Education for International Understanding, the Parliament of the World's Religions, Religions for Peace, and the Asia-Pacific Interfaith Conference and International Peace Research Association (IPRA). He is currently Convener of the Peace Education Commission of IPRA, and a member of the editorial Board of the *Journal of Peace Education*.

Margaret Wade is a lecturer in the Faculty of Health at the University of Plymouth, United Kingdom. She has a background of working as a health visitor and community practice teacher in West Cornwall and has taught in higher education since 1999. Her main teaching area is public health and health promotion in pre- and postregistration nursing programs and she is the Program Lead for the Specialist Community Public Health Nursing program. Currently she is the Chair of the UK Community Practitioner and Health Visitors' Association Education Committee and a member of the National Professional Committee.

Index